Down Under

Mineral Heritage in Australasia

To Harry Conger
with best wishes and kind
regards

[illegible]

Oct. 8, 1992

ISBN 0 646 06721 4

First published in 1992

Designed by Alfred Hickey

Printed in Australia by
Jenkin Buxton Printers Pty Ltd
263 Middleborough Road Box Hill
Victoria Australia 3128

End papers; Paddington Consols gold mine near Kalgoorlie, WA, about 1908.

MONOGRAPH SERIES

1.*	Detrital Heavy Minerals in Natural Accumulates	George Baker	1962
2.	Research in Chemical and Extraction Metallurgy	Ed J. T. Woodcock, A. E. Jenkins and G.M. Willis	1967
3.*	Broken Hill Mines - 1968	Ed M. Radmanovich and J. T. Woodcock	1968
4.*	Economic Geology of New Zealand	Ed G. J. Williams	1974
5.	Economic Geology of Australia and Papua New Guinea - 1. Metals	Ed C. L. Knight	1975
6.*	Economic Geology of Australia and Papua New Guinea - 2. Coal	Ed D. M. Traves and D. King	1975
7.*	Economic Geology of Australia and Papua New Guinea - 3. Petroleum	Ed R. B. Leslie, H. J. Evans and C. L. Knight	1976
8.*	Economic Geology of Australia and Papua New Guinea - 4. Industrial Minerals and Rocks	Ed C. L. Knight	1976
9.	Field Geologists' Manual (Third Edition)	Ed D. A. Berkman	1989
10.*	Mining and Metallurgical Practices in Australasia (The Sir Maurice Mawby Memorial Volume)	Ed J. T. Woodcock	1980
11.	Victoria's Brown Coal - A Huge Fortune in Chancery (The Sir Willis Connolly Memorial Volume)	Ed J. T. Woodcock	1984
12.	Australasian Coal Mining Practice	Ed C. H. Martin	1986
13.	Mineral Deposits of New Zealand	Ed D. Kear	1989
14.	Geology of the Mineral Deposits of Australia and Papua New Guinea	Ed F. Hughes	1990
15.	The Rocks Speak	H. F. King	1989
16.	Hidden Gold - The Central Norseman Story	J. D. Campbell	1990
17.	Geological Aspects of the Discovery of Some Important Mineral Deposits in Australia	Ed. K. R. Glasson and J. H. Rattigan	1990

* Out of Print

Copies of all books currently in print can be obtained from
The Australiasian Institute of Mining and Metallurgy
Clunies Ross House 191 Royal Parade Parkville Victoria Australia 3052
Telephone (03) 347 3166 Facsimile (03) 347 8525

Down Under

Mineral Heritage in Australasia

An illustrated history of Mining and Metallurgy in Australia, New Zealand, Fiji and Papua New Guinea

ARVI PARBO

INCORPORATED BY ROYAL CHARTER 1955

MONOGRAPH 18

Published by
THE AUSTRALASIAN INSTITUTE OF MINING AND METALLURGY
CLUNIES ROSS HOUSE 191 ROYAL PARADE PARKVILLE VICTORIA AUSTRALIA 3052

The generous support of Western Mining Corporation Holdings Limited is gratefully acknowledged

Foreword

In the four countries in which The Australasian Institute of Mining and Metallurgy has branches – Australia, New Zealand, Fiji and Papua New Guinea – the mining and mineral industries have a long and honourable place in national economic and social history.

Australia exported coal before it exported wool. Coal shipments from Newcastle to India and the Cape of Good Hope started in 1800, merino wool shipments from Sydney to England in 1807.

Silver-lead discoveries in the foothills near Adelaide in 1841 started Australia's metal-mining industry and gave newly-arrived settlers hope of escape from their economic stagnation. Copper discoveries elsewhere in South Australia in the 1840s started a series of mines that were to provide employment and export income to generations of Australians for the next 80 years or so.

Prospecting became a popular Australian obsession, and the notion that the continent was rich in minerals gradually gained wide acceptance. It was a notion that proved to be true, and the first officially-recognised discoveries of gold in 1851 began 'the rush that never ended'.

This led to a great surge in Australia's population which gave it that most precious of all resources – people. Most of them did not find much gold, and they turned their talents and energies to other pursuits. Many of our non-mining industries owe their start to people who came in search of gold or to their descendants.

Gold had a certain romance that other minerals have never quite managed to match, but Australia has proven to be one of the world's prime sources of most other minerals.

Australian expertise has played its role in the development of mining industries in Fiji and in Papua New Guinea, in association with local people. New Zealand, too, attracted many people from Australia when gold was first found there. A close relationship between the two countries still continues, with aluminium being smelted in New Zealand from alumina refined in Australia.

The stories of our mining and minerals industries are well worth telling, and I believe this book will help many of our young people come to a better understanding of the social and economic benefits the industry has brought to each of our countries.

J. T. Brady
President – 1991

Dedication

This book is dedicated to all the young people of Australia, in whose hands the future of our mining and mineral heritage rests.

Preface

The discovery and production of minerals have made a tremendous contribution to the betterment of the human life since the beginning of civilisation.

We in the mineral industry in Australasia are very good at what we do. We have been, and are now, amongst the world leaders in our industry. But we have also been, and still are, very bad at explaining what we do, and how this benefits the community.

During my year as President of The Australasian Institute of Mining and Metallurgy in 1990 I spoke during visits to Branches about our mineral heritage. The subject is vast and I was able only to touch on it. This book is based on those addresses.

There is a real challenge to explain to people not familiar with the industry – and this means most people – how the negative attitudes towards mineral developments which have gained such surprising influence in the last twenty years are very much against the community's best interests.

I hope my colleagues in the industry take up this challenge. In today's world it is not sufficient just to do a good job. Informing the public is another vital responsibility of the people in the industry.

Arvi Parbo

Acknowledgements

I am indebted to many people who contributed to the addresses which make up this book. The contributions to individual addresses are acknowledged at the appropriate places in the text. The Deputy Chairman, New Zealand Historic Places Trust, Geoffrey G. Thornton, kindly allowed the inclusion of his paper on New Zealand's Mineral Heritage.

I was privileged to meet many members of The Institute during visits to the Branches and extend my special thanks to all those involved in making the arrangements for those visits. A spirit of professionalism and pride in what they do was very evident.

Gilbert Ralph, MBE, manager of WMC Publications, was instrumental in organising the early visits to the Branches, co-ordinating the preparation of papers, contributing to the research and drafting and mothering the book to its conclusion. Without his help the whole undertaking would not have been possible.

My secretary at WMC, Barbara Giles, made an equally vital contribution through organising the latter part of the visits and preparing many of the drafts. My secretary at BHP, Vicki Barnard, and Gilbert Ralph's secretary, June McWhirter, did the remainder of the drafting.

Vicki Lukins of BHP Public Affairs, with assistance from the staff of BHP archives, very competently prepared a number of addresses, as acknowledged in the text. John Dew, former Editor of *Australian Mining,* made an important contribution to a number of the papers and assisted with editing. Melbourne freelance writer Brian Carroll edited the text for publication and collected the illustrations. John Collins of WMC Publications co-ordinated the printing and publication.

Gavan Collery of WMC Corporate Affairs proof read the copy with thoroughness and Gilbert Ralph and June McWhirter compiled the comprehensive index. Many others assisted by supplying information and illustrations.

I acknowledge the willingness of The Australasian Institute of Mining and Metallurgy to include this book in their Monograph Series and the generous contribution from Western Mining Corporation Holdings Ltd which made possible its publication.

I am grateful to all concerned.

Arvi Parbo

Contents

South Australia

Victoria

Tasmania

New Zealand

Fiji

Papua New Guinea

Past, Present, and Looking to the Future

In a restricted sense mineral heritage refers to the past. It means the discoveries, the developments, the mines and the processing plants, the benefits and the consequences and the human interest stories flowing from these. The history of the minerals industry, inseparable from the history of the countries in which The Institute operates, is a rich background of which we can all be proud.

But there is also a much wider meaning to mineral heritage. Our history has left us with the ability to create and operate world-scale enterprises which can generate wealth for the community. It has left us with the ability to offer satisfying and challenging careers in a wide range of skills to our young people. It has left some of us with the willingness to follow adventurous and challenging vocations at a time when the dreadful mentality of many is to get the most reward for the least effort.

The minerals industry continues to be a vital contributor to improving people's lives. Our mineral heritage lies as much in looking to the future as to the past. It is to this broader context that I now refer.

Coal was the first mineral mined in Australasia, with the first mines at Newcastle in 1797. This 1873 engraving of an undergound scene shows loaded trucks ready to be hauled up the shaft.

The Institute

When the forerunner of The Australasian Institute of Mining and Metallurgy was formed in 1893 in Adelaide, South Australia, the minerals industry in Australia and New Zealand had already been very active for more than 50 years. The first significant mineral discoveries in Papua New Guinea came later, in 1895, and in Fiji even later, in 1931.

The Institute is the association of all the technological professionals in the minerals industry. While the name, for historical reasons, mentions mining and metallurgy only, The Institute in fact includes many other mineral industry disciplines, including geoscientists, engineers, and chemical engineers.

Institute Awards

The Institute is all about people, and in recognition of this there are the Institute Awards to members. They recognise outstanding service to The Institute, the industry, and the profession. The Awards are a part of our mineral heritage.

The conferring of Honorary Fellowships to 'persons of distinction', both within The Institute and outside, is the longest established tradition. To date 69 Honorary Memberships have been awarded.

The Institute Medal, which is the highest Award, is for 'eminent services to the Mineral, Mining, Metallurgical or Petroleum industries'. Fifty such medals have been bestowed since 1935. The annual President's Award, which was introduced in 1974, recognises 'some specific sphere

Surface operations of a typical Newcastle coal mine, 1873.

of a technical nature rather than a managerial role'. There had been 15 President's Awards given to 1990.

More recently established awards include the Sir Willis Connolly Memorial Medal to 'notable communicators', given jointly by The Institute and a group of members calling themselves the Barbarians after the patron saint of German miners, Saint Barbara. There is also the G. B. O'Malley Medal which is the highest award to a student member for a technical paper.

The Award citations and proceedings at award ceremonies, published in the AusIMM's Bulletin, are a valuable biographical record of the recipients.

Prospecting has long been a popular obsession among Australians.

The Past

The Institute's Mineral Heritage Committee, set up in 1984, has the aims of recording and preserving essential features and artefacts of the past, and reminding us all that the present soon becomes the past.

The first Mineral Heritage Seminar was held in Adelaide in March 1986, and its proceedings were published by the South Australian Department of Mines and Energy. The second Seminar was held in Sydney in July 1988.

In 1987 The Institute established the Australasian Mineral Heritage Trust to support:

- research into, and recording of mineral history in Australasia
- preservation, restoration, and identification of sites, equipment, and the like

- research into, evaluation and categorisation of records, and archival data and other information
- establishment, and support for, museums, exhibitions and displays
- encouragement of donations, bequests and legacies.

The activities of the Trust were initiated with the publication, in 1987, of the Leslie Bradford Golden Jubilee Oration.

Both the Committee and the Trust maintain close liaison with bodies such as the Australian Mining Industry Council (AMIC), companies in the industry, Chambers of Mines, the Australian Mineral Foundation (AMF), and the Australian Minerals Industries Research Association (AMIRA). The subject is so vast and the effort needed to make even modest progress so great that the industry cannot afford to duplicate what is being done.

Valuable work in heritage recording and preservation has been done in the past by individual companies and local bodies in various mining centres. One outstanding example is Queenstown, in Tasmania, where a combined effort by the Mount Lyell Mining and Railway Company, the State Government and the Municipality of Lyell in 1983 erected the unique plaza of sculptures by Steve Walker and other exhibits at Miners' Siding to commemorate 100 years of mining in the vicinity of Queenstown. Some other impressive monuments and preserved relics include Morphett's Enginehouse at Burra, the Ivanhoe headframe at Kalgoorlie, miners' cottages at Moonta and mining plant at Ballarat in Australia and the Government Battery at Coromandel in New Zealand.

Limestone quarry plant and trucks, Mount Lyell, Tasmania.

As well as the many physical relics which can be preserved, there are also the written and oral histories. Historical accounts such as those by Professor Geoffrey Blainey, AO, in Australia and Professor John Salmon in New Zealand are invaluable. The stories are about the successes and failures, the marvellous discoveries and the unsuccessful endeavours, those who struck it rich and those who did not.

It is not companies or organisations that do things: it is people. This is where members of The Institute not only can, but must, make a major contribution as a part of their responsibilities as professionals.

Members in positions of authority in mineral industry enterprises have a particular role in this. We should ensure that no company discards or destroys historical records or mining relics without making sure that those of unique value are preserved in a suitable manner. We should encourage the recording of company histories and contribute towards more general works on broader historical aspects of the industry.

Burra. Morphett engine house, winding house and launders, in 1892. Copper discoveries in the 1840s in South Australia gave Australia its first metal-mining industry. Water lifted from the mine flowed along the launders to waterwheels at the crushers and in the open-cut mine.

Much remains to be researched and recorded about the early mineral developments, but there are also more stories waiting to be told about the more recent finds and developments: about iron ore, manganese, aluminium, nickel, copper, uranium, coal, mineral sands, oil and gas and so on. It is essential that they be told. It bears repeating that it is largely the responsibility of members of The Institute to see that this happens.

It is by knowing, respecting, and drawing inspiration from the past that we can cope with the future in a way in which our successors can in due course be proud of our achievements.

Port Pirie waterfront, 1892. Minerals mined and smelted at Broken Hill were exported through Port Pirie from the field's earliest days. In time, all the smelters at Broken Hill closed, and Port Pirie became the largest lead-smelting centre in the world.

The Present

The industry and The Institute are well and active. The President in 1989, Professor Alban Lynch, AM, inaugurated two new Branches, at Roxby Downs and Ok Tedi.

The anti-development and anti-mining campaigns of recent times have slowed down and, at times, deferred mineral projects, but they have not stopped the industry's growth. This is not grounds for complacency. We in the industry have generally not been very good at overcoming the negative attitudes and establishing a positive climate of public opinion ourselves.

We have considered what has been said and come to the conclusion that much of it does not make sense. What does not make sense cannot be pursued far in our professional activities, so we tend to conclude that it would also disappear in the public arena.

This is a serious misunderstanding, because in politics there is no such thing as an automatic rejection of what does not make sense. Great empires have been founded on false ideas. Given enough pressure from particular interest groups, accompanied by skilful media handling, what does not make sense has an excellent chance of becoming law affecting the industry. Much of it has.

Most of the industry's activities take place in remote areas, a long way from the large urban population centres. Most people never see a mine, an ore treatment plant, a smelter, or a refinery except in passing, and do not have first-hand knowledge to test the validity of the misinformation so ably distributed by the opponents. Far from assuming that everybody understands what we do, we must regard it as a part of our professional responsibilities to explain our activities and the industry to the public.

Fluming at Dillmanstown, New Zealand, 1899, showing the Kumaia water race.

Surface plant at Mount Isa Mines Ltd, 1932

Curiously, some of the opposition to today's mineral developments is on the grounds that it would interfere with relics from previous mining, which are now a part of national heritage. At Bendigo in Victoria, a group is proposing to have 40 local government areas, with a population of more than 400,000 people, declared a World Heritage area because it includes the best surviving examples of the 19th century gold rush era in the world. By contrast, when permission is given to new mining projects, the usual conditions are that everything has to be razed to the ground at the end of mining and covered over to resemble the original conditions.

The Institute is probably more conscious than any other group of the importance of preserving our mineral heritage. This can be done sensibly, without excluding areas the size of some European countries from future mining or other economic activity. Preservation of unique features of our past does not require neglect of the present.

A part of our responsibility is to ensure that we and our successors continue to make contributions to the community like those made by our predecessors.

The Future

How can members of The Institute live up to this responsibility?

In its 100 years, The Institute has been a highly-respected body, fostering technical progress and excellence and nurturing professional and ethical standards. It has fostered professional education and the professional standing of its members. We must continue and strengthen all these activities. High professional and public standing and respect are essential to achieving our aims. We must work tirelessly towards this.

But in the changing political and community environment we must also become active participants in public policy formulation affecting our industry and members. Policies are becoming far too important to be left to politicians.

Underground at the South Mine, Broken Hill about 1912 at the time dry fill was used.

Hot strip mill at Port Kembla steelworks. It is one of the world's most advanced.

Gold miners' camp, Western Australia, 1930s.

The key to being able to discharge our responsibilities is in ensuring that we are the premier professional body in the mineral industry. We must attract a high proportion of those eligible for membership as our members.

It is vital to have a high proportion of the eligible students and young graduates as our members. The present becomes the past sooner than we like, and the leadership of The Institute will soon be in the hands of today's students.

The training and education of young men and women to join the ranks of the professions has never been more important than now. We have first-class people today, but the at-times quite defamatory attacks against the industry by small but noisy groups create the danger that other occupations will be seen as offering a higher standing in the community, mainly by those who live in the big cities. The danger is real that the mineral industry may not attract enough of the best young people in the future.

The Institute is doing something about this. The AusIMM Education Endowment Fund set up in 1989 is now operational, and the first scholarships have been awarded. It is focussed on attracting high calibre students to study disciplines relevant to the industry, and on establishing and maintaining contact between the industry and the students during their studies. Company-sponsored scholarships must not stop at simply providing the money. Continuing liaison by the companies with the students is essential, to ensure the greatest benefit for all from the investment and the effort. For scholarships which are not named, The Institute must act in this capacity.

While the Fund is up and running, its effort is limited by the funds subscribed. More financial contributions are needed to build on the good

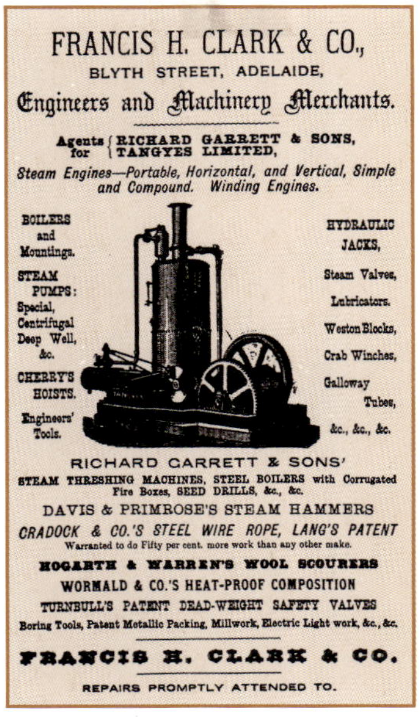

Australia's mineral industry encouraged the development of other industries, such as machinery design, manufacture, sale and maintenance.

Innamincka No. 1. First petroleum well in the Cooper Basin, South Australia, 1959. Its final depth was almost 4 kilometres. Only traces of gas were found and the well was abandoned.

beginning. Mainly for taxation reasons, the Fund's activities are at present limited to Australia. Ways need to be found to extend it to all countries covered by The Institute.

I have already mentioned the importance of taking part in public policy forums. Other organisations – Australian Mining Industry Council, the Chambers of Mines, and so on – represent the industry in this. We represent the technological professionals. Our authority is partly in knowledge, and partly in the number of voices we are able to muster. The leaders of The Institute will speak as often as possible, but we need more voices: the more the better. I urge all members of The Institute to speak up whenever they can, particularly before audiences not familiar with the industry.

We must not think that our individual voices are insignificant and cannot influence the course of events. Public opinion is the most important force in a democratic society. Our individual voices, sounded often enough, and provided there are enough of them, will add up to a powerful chorus.

I conclude as I started: our heritage is as much about the future as about the past. Our industry and professions have the responsibility to continue being great contributors to the community in the future, as they have in the past; this is an essential part of our heritage. There are forces opposing this, which must be overcome as we overcome many other difficulties and adversities in our professional lives. Success in this is largely up to ourselves, the members of The Institute.

This paper was presented at the Annual Conference of The Australasian Institute of Mining and Metallurgy at Rotorua, New Zealand on 19 March 1990.

Aluminium ingots at the Alcoa smelter, Point Henry, near Geelong, Victoria, in the 1970s.

New South Wales

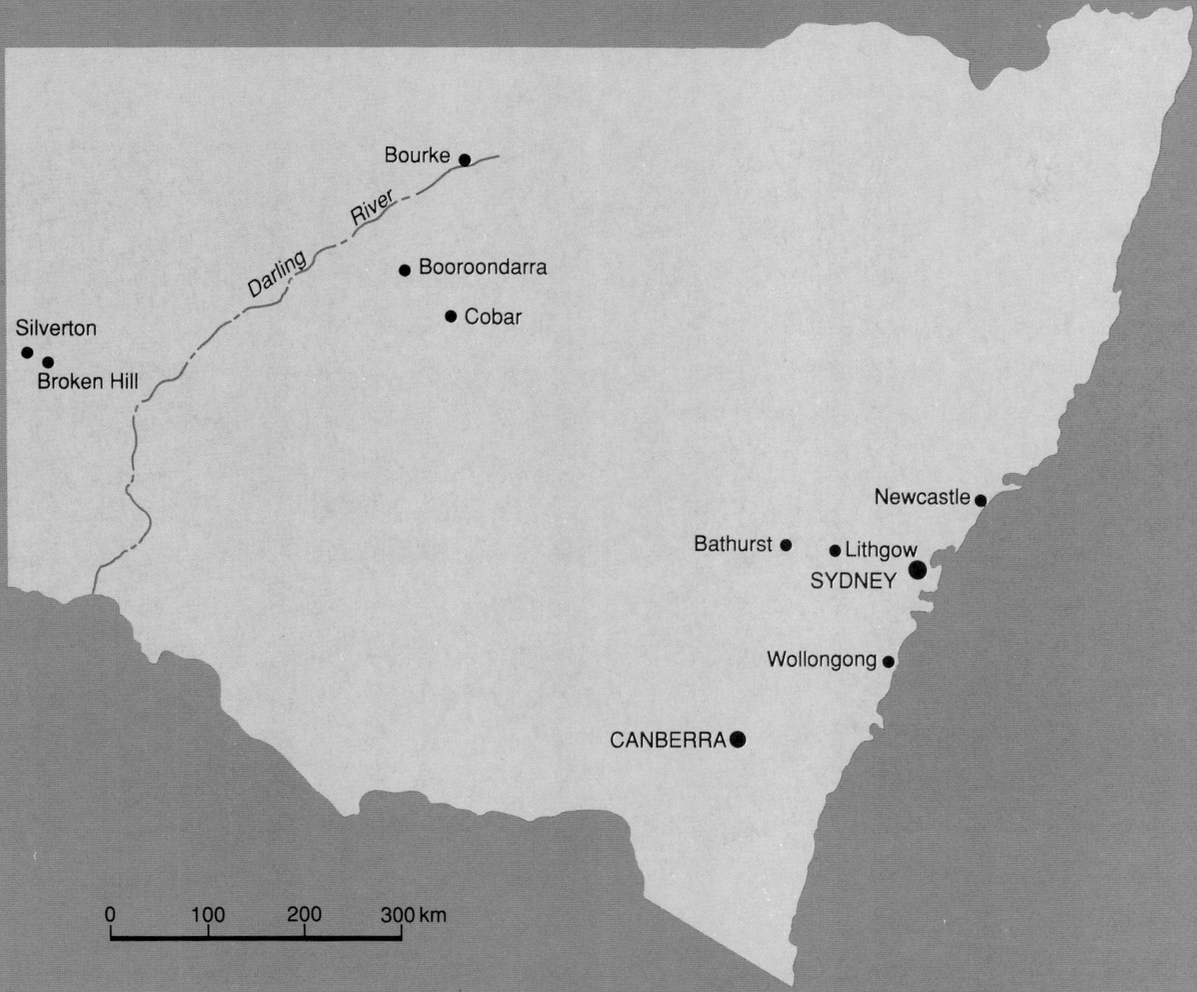

Minerals in the Economic Development of Australia

Australia's mineral industry began with coalmining at Newcastle in 1797, as this 1947 stamp records. Australia exported coal before it exported wool.

In 1797, when the infant colony of New South Wales was nine years old, Lieutenant John Shortland, searching for runaway convicts north of Sydney, found outcrops of coal by the mouth of a river, later named the Hunter.

A few convicts, one of whom had been a miner, were sent to the site and a small settlement known as Coal Harbour was formed. Australia's export income from minerals began in 1800, when the master of the ship *Cornwallis* bought 45 tons of coal. In 1804 the place was given the name Newcastle. It was destined to become one of Australia's great regional and industrial centres.

Economic development and population growth based on mineral discovery and production were to be repeated many times throughout Australia, right through to the present. There is major potential for similar development and growth to continue.

Australian metal mining was born early in 1841, when Thomas and Hutchins, immigrant Cornish miners, fired grass at Glen Osmond, in the foothills south east of Adelaide, and uncovered an outcrop of galena. The discovery was worked in a primitive way until 1847 when, as would so often happen in our early mining history, a London company was formed to finance systematic working. Lead from furnaces at The Glen and in Adelaide was being shipped to England by 1848, but the deposits were small, and the mine was deserted by 1851. Mines set up to work finds of lead and copper near Geraldton in Western Australia in 1842 were also short lived.

The first major metal mining in Australia was for copper, at Kapunda and Burra, north of Adelaide. Pastoralists discovered the Kapunda deposit late in 1842. At the Burra deposit, found in 1845 by a shepherd, early output averaged 20 per cent copper. Export values of copper ores and concentrates soon topped those of South Australia's wool and wheat, bringing new confidence to a young colony emerging from depression, and attracting migrants needed for its development. At least 5,000 people lived in its copper-mining areas by 1851.

A relic of Australia's first metal mine. Chimney at former smelter at Glen Osmond, near Adelaide, where a silver-lead deposit was found in 1841.

Similar benefits came on a spectacular scale with the discovery of gold in New South Wales and Victoria in 1851. On 5 May 1851 the *Sydney Morning Herald* reported: 'It is no longer any secret that gold has been found in the western country. The fact was first established on the 12 February by Mr E. H. Hargraves who returned from California a few months since'. By 15 May 300 diggers were at the site, near Bathurst, which Hargraves, with a fine sense for publicity, had named Ophir, after the Biblical city of gold.

Other finds were to be made in New South Wales, but those in Victoria far outdid them. A discovery at Clunes was announced in 1851, and finds at Ballarat, Mount Alexander, Daylesford, Bendigo and Beechworth quickly followed. The news crossed the oceans. People, rich and poor, farmers and peasants, shop-owners and factory workers, remittance men and clergy, flocked to the new El Dorado. They came from all points of the compass: from Europe, Asia, Africa and the Americas. The impact was dramatic. 'Numbers continuing to arrive are extraordinary – 1000 a day', wrote Melbourne merchant James Graham on 22 January 1852. 'The day before yesterday I counted no fewer than 11 sails in sight, all inward bound – publicans and shopkeepers are making a fortune'.

Australia's population was 405,000 at the time of the discovery of gold. By 1860 it would be more than 1,000,000. Companies and machinery had by then displaced the individual digger as the main form of mining. There was a flurry of mining floats. J. B. Were and Associates formed Melbourne's first Stock Exchange in May 1860. Sydney would follow in 1870.

In his great pioneering work on our economic history, *Labour and Industry in Australia*, T. A. Coghlan assessed that the first decade of gold '. . . saw a revolutionary change in the relations between employers and employed, the destruction of the last vestige of convictism and, industrially, the new creation of Australia'. Decades before Federation in 1901, W. C. Wentworth, elder statesman of New South Wales, observed that 'gold precipitated Australia into nationhood'.

To the gleam of gold was added in 1859 the warm glow of more South Australian copper, found at Wallaroo on Walter Hughes' sheep station.

Australia's development entered a new era in 1851 with the discovery of gold at Summer Hill Creek, near Bathurst, by Edward Hammond Hargraves. People from all over the world joined the gold rushes and boosted Australia's population.

Discovery of the nearby Moonta mine, the first mine to pay £1,000,000 in dividends, quickly followed, and 36 furnaces were at work at Wallaroo by 1866. An initial gift of £20,000 by Hughes and later £100,000 by his partner Sir Thomas Elder, GCMG, would lead to the founding of the University of Adelaide.

In 1867, discovery of gold at Gympie started the first of the Queensland rushes, and by the mid-1880s more gold finds followed the coastal ranges far north to the Palmer River. Mount Morgan was among the most famous. Discovered in 1870 by the stockman William Mackinlay, it was taken up in 1882 by a syndicate and floated as Mount Morgan Gold Mining Co. Ltd in 1886. Rich outcrops yielded as much as 10 ounces of gold to the ton, with underground ore averaging four ounces per ton.

Mining exhibit at the 1880 International Exhibition in Melbourne.

The immigrant William Knox D'Arcy was one of the original syndicate. After he went back to London, wealth from Mount Morgan financed his expenditure of almost a quarter of a million pounds in searching for oil in Persia. The search succeeded and led to the formation of the company that later became British Petroleum Ltd.

Tin was found in 1871 at Tingha, near Inverell in northern New South Wales. In 1873 tin deposits were being mined over 160 kilometres of country through to Stanthorpe in southern Queensland. John Moffat built a tin smelter and a model town at Tent Hill and developed the later field near Herberton, in northern Queensland. His money and energy developed the Chillagoe copper-silver-lead field, and Mount Elliott near Cloncurry. He built railways and dams, and much of the wealth he won from minerals he gave to charities.

At about the same time, the world's greatest tin mine until then was discovered at Mount Bischoff, in north-west Tasmania. Found in 1871, the mine had paid £210,000 in dividends by 1880. Production to 1888, along with that of mainland mines, made Australia at that time the world's largest tin producer.

Tasmania's greatest mineral discovery would be the copper and gold ore bodies of Mount Lyell, found in 1883. Smelting problems bedevilled the operations until in 1896 the German metallurgist Robert Carl Sticht succeeded in using sulphur in the ore as fuel in the process. At Cobar, 570 kilometres west of Sydney, copper had been mined as early as 1871, but costs in the remote inland area were to delay large-scale production for many decades.

Benefits from mining continued to radiate through the economy. Melbourne's incredible growth was a typical example with the International Exhibition in 1880 hailed as a magnificent Australian achievement. Yet everything that had happened so far was eclipsed by the glitter of silver, lead and zinc in the Barrier Ranges, in the far west of New South Wales, near the 'broken hill' Charles Sturt had noted in his 1844 explorations.

Boundary rider Charles Rasp made the discovery in 1883, and in 1885 a syndicate floated The Broken Hill Proprietary Company Ltd (BHP).

The oxidised surface ore had hundreds of ounces of silver to the ton. It soon gave way to deeper sulphide ore containing silver, lead, and zinc.

Tipping copper ore for haulage to the surface at the Wallaroo mine, Kadina, South Australia. Copper mining has been a major component of the South Australian economy since the first discoveries at Kapunda in 1842.

Separation of the metals, at first a problem, was achieved by the invention of flotation. Pioneered in Australia, flotation revolutionised treatment of sulphide ores throughout the world. Charles Vincent Potter, a Melbourne brewer, took out the first patent in 1901. BHP's General Manager, Guillaume Delprat, CBE, discovered a similar process in 1902 and the two were merged in 1907. Meanwhile Auguste de Bavay, a Melbourne brewery chemist, developed his skin or film flotation process. Flotation technology was developed further to treat the zinc rich tailings which had been discarded at Broken Hill, separate several sulphide minerals in the same ore, and recover non-metallic minerals.

As the great mines of Broken Hill developed, a spirit of optimism developed even faster, with Melbourne and Sydney the focal points for capital formation and investment. In Melbourne, a building boom peaked in 1888, a year in which imports to Victoria were £24,000,000, almost double the value of exports. Over-extended investment plummeted with the bank crash of the early 1890s, and Australia was in a depression.

Then, in 1893, came the great gold discoveries at Coolgardie and Kalgoorlie in Western Australia. Bayley's Reward, Lake View Consols, Golden Horseshoe, Ivanhoe, Great Boulder: these were some of the mines which rocketed Western Australia to leadership of Australia's gold industry. In 1898-99, Lake View was the biggest gold producer in the world.

Soon after, hills of iron were replacing the silver hill as BHP's centre of interest. BHP's Broken Hill mine was at the apex of the great Broken Hill orebody. The ore dipped away into the leases of other companies at either end, but the BHP mine had a limited life. The company obtained

Early days in the open-cut copper mine at Mount Morgan, one of the longest-lived and richest in Queensland.

leases over iron ore deposits at Iron Knob, 53 kilometres inland from what is now Whyalla, in 1897. The ore was at first mined for flux for the lead smelters at Port Pirie, on the other side of Spencer Gulf, but it had greater potential.

Visiting Iron Knob in 1900, Harvey Patterson, then Chairman of BHP, observed that 'such an enormous quantity of ore is exposed that I see no reason why it should not supply, if required, and considered advantageous, a portion of the present requirement of the world for the manufacture of iron'. Guillaume Delprat agreed. He arranged a trial smelting of 5 tons of pig iron at Port Pirie in 1907. Declared 'excellent' in Adelaide, it sold at £5 a ton.

Remains of Tasmania's richest gold mine, at Beaconsfield, on the Tamar River. Gold was first found there in 1847, with a major strike in 1877. The mine was down to 450 metres when pumping out the inflowing water became too costly. It closed in 1913.

The Hoskins Brothers were already producing iron and steel in Lithgow, New South Wales, but output there was less than three per cent of Australia's demand. The proposed construction of the east-west trans-continental railway would require more steel. On 12 June 1911 BHP's directors decided to consider making iron and steel.

David Baker, mining engineer and consultant, of Philadelphia, arrived to plan and supervise construction of the works. He was enthused by the quality of the Iron Knob ore.

Newcastle, the place of the first production and sale of an Australian mineral in 1800, became the site BHP chose to build its first steelworks. Construction began in April 1913, and the blast furnace was blown in on 9 March 1915. Australia was by then involved in the First World War. With the interruption of steel supplies from overseas, most of Newcastle's growing output was needed for domestic use.

In 1921 Essington Lewis, CH, took over from Delprat as General Manager. Over the next two decades the initial steel making venture grew into a massive industrial complex, integrated vertically and horizontally. It became so efficient that at the outbreak of the Second World War in 1939, BHP was producing the world's cheapest steel.

With the decision to enter iron and steel, BHP in 1915 sold a majority interest in the Port Pirie lead smelters to a consortium of North Broken Hill Ltd, Broken Hill South Ltd and Zinc Corporation Ltd, which formed The Broken Hill Associated Smelters Pty Ltd (BHAS).

W. L. Baillieu and W. S. Robinson were the major figures behind the Melbourne-based Collins House group, as their companies came to be known. In the 1920s the group created Electrolytic Zinc Company of Australasia Ltd (EZ), British Australian Lead Manufacturers Ltd (BALM), Metal Manufactures Ltd, and the Electrolytic Refining and Smelting Company of Australia Ltd (ER&S). It took park in setting up Imperial Chemical Industries of Australia and New Zealand Ltd (ICI). In 1936 the group formed Associated Pulp and Paper Mills Ltd (APPM)

Workmen open the taphole of No. 1 blast furnace at the BHP steelworks at Newcastle. BHP made steel for the first time at Newcastle on 9 March 1915.

Miners about to go underground at Boulder Perseverance gold mine, on Kalgoorlie's Golden Mile, 1934. Discovery of gold in the Coolgardie-Kalgoorlie era was one of the few bright spots in the economy of any of the Australian colonies in the 1890s and led to a surge in Western Australia's population. In the depressed 1930s, goldmining again became a source of work to thousands of people.

to make writing and printing paper from Australian hardwoods at Burnie, Tasmania. It also joined with BHP, General Motors-Holden's Ltd (GMH) and the Orient Steam Navigation Company Ltd to form the Commonwealth Aircraft Corporation Pty Ltd (CAC) to produce aircraft and aircraft engines at Fishermans Bend, Melbourne. The genesis of all these economic developments was the discovery of the minerals at Broken Hill.

Meanwhile, soon after the First World War, Victoria's newly-formed State Electricity Commission adopted German technology for using brown coal to make briquettes and generate electricity. Sir John Monash, GCMG, KCB, one of Australia's most famous soldiers and an experienced engineer, returned to head the Commission. Under his chairmanship, billions of tons of brown coal in the Latrobe Valley set Victoria on the way to meeting all its power needs and more, bringing widespread economic development.

In the 1920s, low world prices caused a downturn in metal mining. In this bleak context the massive copper-lead-zinc mineralisation of Mount Isa was discovered in 1923. Against great difficulties, William H. Corbould of Mount Isa Mines Ltd (MIM) was its champion. A world-wide economic depression cut the flow of capital from London, but in 1930 the American Smelting and Refining Company Ltd (ASARCO) rescued MIM with an injection of funds. The first small profit was made in 1936-37, but the company did not pay a dividend until 1947. In the decades that followed, MIM became one of the great lead-zinc-copper-silver producers of the world.

Inside the cell room at Risdon electrolytic zinc refinery, in the 1950s. The use of electricity for minerals treatment was a major factor in the steady development of hydro-electric power and the Tasmanian economy.

The picture was much brighter with gold. In 1931 came a long-awaited rise in the world price of gold, more than doubling it to £8.10.0 an ounce, including the effect of depreciation of the Australian currency. A gold mining revival followed. Claude de Bernales floated many Western Australian mining ventures which did not have long lives, but provided much-needed employment to many. The work of W. S. Robinson was more enduring. In 1930, anticipating the price rise, he influenced British, South African, and Australian interests to form a number of companies to explore for and produce gold in Australia. Amongst these, in March 1933, was Western Mining Corporation Limited, which became mainly Australian-owned and the major company in the Group in 1949.

'I am far from sure that the public, or even Governments, recognise how much of the recent recovery in Australia is because of the renewed activity in gold mining', WMC Chairman Colin (later Sir Colin) Fraser said in March 1935. There was a return of confidence and employment in a wide range of industries.

During the Second World War the Federal Government provided £2,000,000 in a crash programme of exploration towards production of additional strategic metals, such as tungsten and aluminium. The only significant result from this was that the scheelite deposits of King Island in Bass Strait were demonstrated to be of world class.

Looking to the future of Australia, W. S. Robinson focused on the need to develop the Northern Territory and north Queensland. Writing to his son L. B. Robinson on 11 August 1948, he considered reliance on development of the cattle industry, but rejected it. 'Australia's experience shows that above all gold has been the great populator', he wrote. 'It is the miner who is prepared to face risks of climate, of the absence of houses, sanitation, water, electricity, or other lighting and power services. He is the cheapest of all settlers, he pays for himself and takes the risks'.

Interest in uranium for military purposes triggered a uranium exploration boom in the late 1940s and early 1950s. This was followed by the demand for uranium for power generation, commencing in the mid-1950s. The first significant uranium discovery in Australia was at Rum Jungle, in the Northern Territory, in 1949. This was followed by developments at Radium Hill in South Australia and in Queensland in the first half of the 1950s.

Sinking a new shaft at Mount Isa Mines, 1971. Discovered in 1923, the Mount Isa field has been developed steadily and is a major factor in the economy of north Queensland.

Until the late 1950s, commercial bauxite was not known in Australia. For strategic reasons, the Commonwealth Government sponsored small-scale alumina and aluminium metal production from imported bauxite at Bell Bay in Tasmania, where production began in 1955.

In the 1950s finds of huge deposits of bauxite at Gove, and particularly Weipa, both on the shores of the Gulf of Carpentaria, and in the Darling Range in Western Australia became the basis of a world-scale Australian aluminium industry. Alumina refineries were built in the Northern Territory, Queensland and Western Australia, with smelters in Victoria, Queensland and New South Wales. Australia became the world's largest producer of bauxite and alumina, a substantial producer of aluminium metal, and a major exporter of all three.

Very large iron ore deposits discovered in the Pilbara area of Western Australia in the 1960s became the basis of a world-ranking iron ore export industry, at first to Japan, later also to other Asian countries and Europe. Circumstances had changed since W. S. Robinson's comments on the modest demands of miners; modern towns with first class facilities and services were provided by the mining companies in what had been virtually uninhabited areas. The mining companies built some of the world's most efficient railways to transport the ore to newly-established harbours and shiploading facilities.

In 1966, discovery of the nickel deposits at Kambalda triggered a major exploration and mining investment surge. Money flowed into Australia from overseas, mainly from the United States, Britain and Japan. The London *Financial Times* declared Australia to be the healthiest economy in the world. Contracts were signed for the export of manganese ore from Groote Eylandt. Australia became the world's major supplier of mineral sands. Drilling offshore in the Bass Strait discovered large reserves of oil and natural gas. Other oil and gas finds were made at Moonie, in Queensland, and near Barrow Island, in Western Australia. By 1969 Melbourne, Brisbane, and Adelaide were all supplied with natural gas. Australia could now meet 60 per cent of its oil and gas needs from local sources.

In Queensland, discoveries of enormous coal deposits in the Bowen Basin in the 1960s gave rise to a major export industry of coking and, later, steaming coal. Some of the steaming coal, used in power generation, encouraged the setting up of an integrated aluminium industry at Gladstone.

Off the coast of Western Australia, in an area known as the North West Shelf, large natural gas fields were discovered and became the source of gas for Perth and for a major liquefied natural gas (LNG) export industry. Exports of North West Shelf LNG to Japan began in 1989. Further drilling indicated potential for oil production as well.

Additional uranium deposits were discovered in the Northern Territory, and two were brought into production. In 1975 the discovery hole was drilled at the Olympic Dam copper-uranium-gold deposit in the north of South Australia, where production began in 1988 from a very large underground orebody. Oil discoveries in the Timor Sea came in the 1980s, just as production from Bass Strait started to decline.

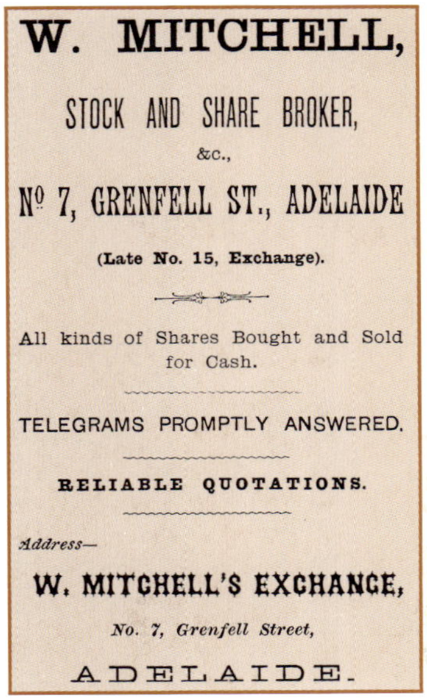

Australia's stock and share brokers have had a long involvement in the mining and mineral industry. This advertisement appeared in 1888.

Gold mining in Australia had another boom in the 1980s, because of good prices, a favourable taxation regime, and advances in technology which made it possible to treat low-grade and near-surface oxidised ores. Production of gold increased from 17 tonnes in 1980 to an estimated 225 tonnes in 1990. Without this extra export income, Australia's current account would have been in a far sorrier state.

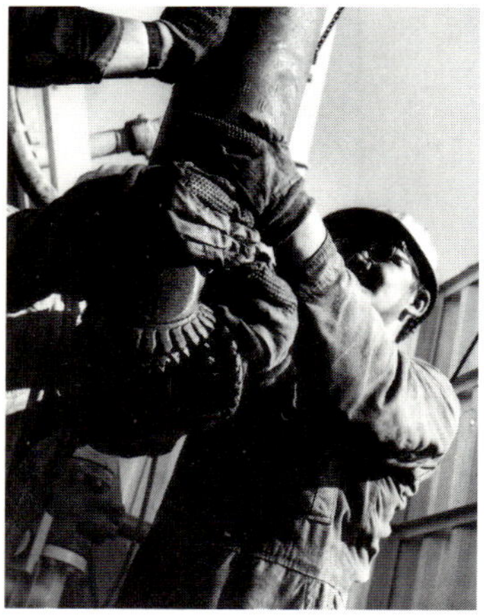

Drill on an Esso-BHP oil rig in Bass Strait. Discovery of gas and oil there in 1965 added a major new industry to the Victorian economy and relieved Australia's balance of payments of the need to import much of the nation's oil needs.

I have had to rush through the major mineral developments in the last 190 years, without mentioning many others. For the last 150 years from the South Australian copper finds in the 1840s, the mineral industry was absolutely vital to the development of Australia and the well-being of Australians. Time and time again, it came to the rescue when Australia faced grave economic circumstances.

It remains so: at a time when the deficit on the external account is Australia's gravest economic problem, the value of mineral industry exports exceeds $16 billion per annum or 41 per cent of all export income, close to 50 per cent if processed minerals listed under 'manufactures' are taken into account. The industry employs directly 170,000 people and since 1967 has built 25 new towns, 12 new ports, 20 airfields, and almost 2,000 kilometres of railways.

The record of the industry speaks for itself. It is up to industry people to make sure that this record is understood and appreciated by the community, and that the industry can continue making this vital contribution to Australia. It is in the interests of every Australian that it should do so. We need the mineral industry now as much as ever.

I am indebted to Frank Strahan, University of Melbourne, Archivist, for his invaluable assistance in the preparation of this paper which was presented in Sydney on 20 August 1990.

Mineral exports are a vital source of export income to Australia. More than 100 vessels a year load alumina at the Kwinana jetty for smelter customers around the world. The alumina is refined from bauxite mined in the Darling Range, not far inland from Perth.

Australia's Mineral Industry Born at Newcastle

Australia's earliest white settlers soon began to make use of the continent's mineral wealth. The 'industry' began in 1797, when small quantities of coal were taken from Newcastle to Sydney.

A settlement, Coal Harbour, sprang up at the mouth of the (then) Coal River – now the Hunter – and organised mining (probably of the Dudley seam) began when Governor Hunter sent 16 convicts there in 1801. Coal was cut by hand and carried in baskets by convicts over the coal carriers' track. At first there was no wharf, so they waded out to the small sailing vessels.

When a wharf was built, the improvements amounted to hand-carts, bullock wagons and wheelbarrows, the ready supply of convict labour being a strong disincentive to mechanise. To the observer the scene at the wharf was bleak. One journalist wrote: 'Coal is doled out in miserable thimble-fulls and drawn about the wharf in hand barrows'. Newcastle was 'a dirty infant of a town'.

Lieutenant John Shortland, who found coal on the banks of a river he discovered north of Sydney in 1797.

Production methods changed when the Australian Agricultural Company took over the mines in 1828. Coal production, 4,000 tons in 1830, rose to 34,000 tons in 1842. Convict labour was being phased out and the company began to bring in free miners from Wales and Ireland.

The Australian Agricultural Company brought with it a different attitude to mechanisation, at least for coal handling. In November 1831, it advertised that it was ready to supply coal at its new wharf and on 10 December the coastal steamship *Sophia Jane* took on two skips of coal, the first to be loaded by mechanical means.

The industry grew swiftly from the late 1850s. Demand in Sydney and Melbourne boomed, and a vigorous export trade to America, New Zealand and the Pacific Isles grew up. The Australian Agricultural Company's monopoly was broken as others entered the industry. Handling methods improved with the use of cranes and a standard wagon (of three tons capacity) designed to discharge the bulk coal into ships.

A further development was the removable coal hopper wagon (with a capacity of six tons). Based on an original concept developed by the Scottish Australian Mining Company, it was introduced in 1863 with the opening of the Lambton Colliery.

Newcastle's population, 8,000 in 1861, increased rapidly, and associated developments (such as banks opening branches in the town) took place. Other townships sprang up near the mines.

The Australian Agricultural Company's mine at Newcastle. The company took over the government mine in 1828, but still used convict labour to work it.

Related Developments in the 19th Century

During the 19th century Newcastle, premier coalfield in the Australian colonies, yielded over 60,000,000 tons. Then as now the coal industry underpinned major industrial developments, as well as earning export

Shipping in Newcastle Harbour, about 1850. As more steamships appeared in Australian coastal waters, demand for coal grew steadily.

income. Among the most significant of the industrial developments were steam power, gas production and smelting.

Copper smelters at Wallaroo in South Australia, for instance, used about 10 per cent of all coal shipped from Newcastle in the 1860s. From 1867, the returning colliers carried lower-grade ore as ballast back to Newcastle, where the Hunter River Copper Works operated for a quarter of a century. Another significant development was the world's first electrolytic zinc plant, at Cockle Creek, in 1896. Although the plant had little success, its lead smelter continued to operate, and Sulphide Corporation made a great contribution to the Hunter.

By the late 19th century Newcastle's coal industry faced strong competition: externally from Japan and Canada as well as from Britain, and closer to home from the southern coalfields (notably after their rail link with Sydney), and with the opening of mines in Victoria and Queensland. Even so, between 1886 and 1894 Newcastle exported 6,000,000 tons of coal, much of it still by sailing vessels rather than steamships, and New South Wales was said to have the most extensive coal handling equipment in the southern hemisphere. Growth in the industry was reflected in population: at Federation, in 1901, Newcastle had over 50,000 people.

Just after the turn of the century, New South Wales coal exports had risen to 2,500,000 tons per annum, but a major strike in 1909 caused a serious setback. After 1924-25 exports to other states and overseas declined markedly, hard hit by shipping disputes and rising prices. By 1945 overseas trade had virtually ceased. According to C. M. MacDonald (then Chairman of the Northern Colliers' Association) the workers 'did their best deliberately to murder (the industry)'.

Fortunately, just as coal mining was on the wane, the coalfields provided the means for another major industry to take over as the economic mainstay of the Hunter.

Newcastle coal miner, hand boring at the coal face, Northumberland Colliery, about 1900.

Birth of the Steel Industry

In the early years of this century The Broken Hill Proprietary Company Ltd (BHP), under the leadership of Guillaume Delprat, began to think of setting up a steel industry in Australia. A small works existed at Lithgow, but BHP had in mind a much larger investment. Australia was importing a million tons of steel every year, at a cost of £6,000,000.

At the time, three tons of coal were needed to smelt a ton of iron ore. Steelworks therefore needed deep-water frontages close to a source of coking coal. David Baker, the American expert employed by BHP in 1912 to explore the feasibility of a steelworks, suggested the choice was between Newcastle and Port Kembla. After extensive testing he concluded coal from the Bore Hole seam at Newcastle would make good blast furnace coke because 'it would carry a good burden and work kindly in the furnace'. In 1912 the decision to build a steelworks in Newcastle was announced and the New South Wales Parliament passed the Newcastle Iron and Steelworks Act.

David Baker, the United States consultant BHP engaged to advise it on selecting a site for and developing its proposed steel industry.

The site for the steelworks was that of the old Wallaroo Company's copperworks. It was swampy, and partly under water at high tide. It was drained and reclaimed and after what today seems a remarkably short time, the works were completed. The blast furnace was blown in on schedule in March 1915, the first blooms and billets were rolled on 9 April and on 24 April, eve of the Gallipoli landing by Australian soldiers, the first steel rails were made. The steelworks, with a capacity of 150,000 tons per annum, was officially opened on 2 June.

The works were extended to cope with the increased demands of wartime and before the war ended production capacity had almost doubled. Iron ore from Iron Knob and Iron Monarch in South Australia, together with Newcastle's coal, were underpinning the supply of steel rails to the Western Front in Europe as well as Australia. Shell steel was being supplied to Australian manufacturers, while 30,000 tons of steel products were exported to Britain, and Newcastle began to roll plates for shipbuilding in Australia.

Wartime needs stimulated other development. New heavy industries came to Newcastle and old ones, such as the engineering firm Goninan & Company, expanded. Activity was not restricted to the Hunter. The growth of the steel industry required further developments at Whyalla and Iron Knob. Limestone quarries, in New South Wales and Tasmania, were enlarged or developed to meet the demands of the industry. Local supplies of magnesite and fluorspar also had to be developed.

Jetty built for construction of the Newcastle steelworks, 1913. The site, formerly that of the Wallaroo company's copper works, was swampy and partly under water at high tide. It had to be drained and reclaimed.

Hard times : Good times

After the war, the Newcastle steelworks suffered strong international competition. Between 1915 and 1920 coal prices in Australia had trebled and wages had almost doubled. BHP's costs were double those of British steelmakers, which made its product uncompetitive. Professor Geoffrey Blainey, AO, has suggested that coal mining and coastal shipping were at the time 'possibly two of the most inefficient industries in the country'. When efforts to reduce costs came to nothing, the steelworks closed in May 1922 and stayed shut for nine months.

Bloom mill operators, Newcastle steelworks, 1915.

Advertisement for BHP steelworks in an export directory, 1922.

The steelworks resumed production in 1923-24, with an annual output close to 290,000 tons. During the Depression of the 1930s this was slashed and Newcastle, then the sixth-largest city in Australia, suffered greatly, with about one third of her breadwinners out of work.

Between the wars, the iron and steel industry expanded further to include finished products, with companies such as John Lysaght Ltd and Stewarts & Lloyds Pty Ltd opening up in Newcastle. By the mid-1930s the steelworks were working at a hitherto unknown rate and BHP steel was the lowest-priced in the world. At the outbreak of the Second World War, Newcastle steelworks had increased its capacity to 1,000,000 tons and was the largest integrated steelworks in the British Empire. The war years once again focused people's efforts and led to many products being made for the first time.

Postwar Development

After the Second World War, significant expansion took place in the 1950s and 1960s, including major technical developments. The Basic Oxygen Steelmaking (BOS) process was adopted in 1962, and the works computerised.

A downturn in both domestic and international demand in the 1970s and early 1980s brought with it real hardships. The Steel Industry Plan, put in place in 1983, focused on efficiency and quality, and made a strong recovery possible. The industry, which still largely determined the well-being of the Newcastle region, was placed on a sound footing once again.

Steel was featured on this stamp, issued in 1947 to mark Newcastle's 150 years.

Revitalising the Coal Industry

During the 19th century various innovations improved coal handling and the efficiency of the port, but coal mining methods saw few changes. Not until 1935 was coal production by mechanical means seriously attempted in Australia. In that year, Lambton Colliery became Australia's first completely mechanised colliery, equipped with heavy track-mounted coal cutters, coal-boring machines, track-mounted coal loaders, battery locomotives, and large capacity skips. Other mines followed this lead. Australian-made equipment featured in some of the mechanisation programmes when the Second World War interrupted supplies from the United States.

Another important technical development was the use of conveyor belts in cross measure drifts. Pioneered in Australia at Lambton Colliery, at the time only the second mine in the world to use it, this technique superseded the vertical shaft hoisting system.

Coal exports from Australia had a resurgence from about 1952. Between 1952 and 1962, exports from Newcastle grew from 54,000 to

Newcastle steelworks, 1990.

Since the Second World War, machinery has taken over the task of hewing coal.

1,500,000 tons per annum. Port facilities were progressively upgraded and longwall mining techniques adapted to cope with local geological constraints.

In 1988-89, total coal production in the region amounted to 42,500,000 tonnes, approximately 70 per cent of it exported. Valued at A$1.28 billion in 1989, coal exports through Newcastle represented about 30 per cent of Australia's total coal exports.

Moving iron from the blast furnaces to the steelmaking department by the trainload, each railcar with 200-ton capacity, in the 1970s.

Other Major Industries

Besides iron and steel, other major industries in the Hunter Region were established or expanded in the post-war period, many of them relying on the coalfields. Most of the coal not exported was used for electricity generation, the Hunter Region producing almost 80 per cent of the state's electricity, some of it used in the local aluminium and lead-zinc industries.

Newcastle's lead-zinc industry pre-dates the iron and steel industry. In 1990, works at Cockle Creek, now owned by Pasminco Ltd, had an annual capacity of over 80,000 tonnes of zinc. The plant uses the Imperial Smelting process which, when introduced there in 1959, enabled substantial cost reductions. Aluminum smelters at Kurri Kurri (established in 1969), and Tomago (1984) together accounted for about 30 per cent of Australia's output of aluminium metal.

Mineral sands mining also had a long history in the region. Production topped 115,000 tonnes in 1987-88, for an ex-mine value of almost $49,000,000. Industrial sands, brick and refractory clays were also important products.

Bayswater Power Station. Modern power stations are built on the coalfields, often right next to the mines that fuel them.

Future Outlook

Over the years the Hunter Region diversified well beyond its mining and steel industry base. By 1990 the tertiary sector was the largest source of employment there. This was as it should be. Unlike so many other mining areas, the Hunter was not hampered by extreme isolation or a harsh climate which had constrained development elsewhere in Australia. We could expect the trend towards diversification to continue. Even so, the region's mineral heritage would continue to exert a powerful influence on life in the Hunter.

The New South Wales Coal Association was confident about the outlook for the industry, based on the increasing use of the PCI (pulverised coal

Syphoning molten aluminium from one of the pots in Alcan's aluminium smelter at Kurri Kurri, in the Hunter Valley.

injection) process in steelmaking and, given the low sulphur content of the region's coking coal, its relative environmental acceptability. The coal industry had a number of projects under construction, as did the Electricity Commission of New South Wales.

BHP was building an electrolytic manganese dioxide plant in Newcastle, an investment of some $100,000,000, using a manufacturing process developed at the company's Central Research Laboratories in Shortland. It was a sign of the times that there were plans for part of the John Darling Colliery site, which ceased production in 1987, to be turned into a tourist resort, incorporating a large flora and fauna reserve. Pasminco Ltd's Cockle Creek plant had embarked on a $50,000,000 capital expenditure programme, which was aimed to increase productivity by 15 per cent.

As one of our older and more mature mining and industrial areas approaches the next century, its mineral heritage continues to provide for the well-being of large numbers of Australians.

Zinc stockpile at Sulphide Corporation Pty Limited, Boolaroo, in the 1970s.

I am indebted to Vicki Lukins for her major contribution to this address and to Peter Scaife of BHP Research Centre, Newcastle, John Dew and Gilbert Ralph, MBE, for their assistance in the preparation of the paper, which was presented in Newcastle on 16 July 1990.

Since the Second World War, Newcastle steelworks has been steadily upgraded to keep it abreast of developments in technology. Much of the work became computer controlled.

Coal and Steel from the Illawarra

The changing fortunes firstly of the coal industry and then the steel industry have dominated life in the coastal strip south of Sydney known as the Illawarra. Great achievements have been tempered by major hardships. Those with resilience and ingenuity have been able to find and seize upon opportunities.

Coal Mines and Coke Ovens

This pattern began early on. Shipwrecked sailors first found coal in the Illawarra in 1797. George Bass, sent by Governor Hunter to verify the discovery, reported finding a stratum of coal '6 feet deep in the face of a steep cliff which was traced for 8 miles in length'. Despite this, the first serious attempt to work the field did not come for another half a century.

Instead, the region became one of the colony's richest dairying areas, and the small port of Wollongong sprang up to support this activity. By 1826, the town boasted a population of forty-six.

Eventually, in 1849, James Shoobert opened a shaft in the coal seam on the slopes of Mount Keira. Operations were on a small scale, but this did not detract from the excitement felt when the first load of Mount Keira coal entered Wollongong for loading on a steamer:

> This was the occasion of a procession comprising a band, miners, pedestrians (a goodly number) – the coal carts, each drawn by two horses and horsemen – headed by Messrs. James Shoobert and W. D. Meares J.P., the former gentleman being the lucky proprietor of the immense coal mountain.

Despite the fanfare, Shoobert's mine was not very successful. The first commercial quantities of Illawarra coal were produced in 1857, when Henry Osborne began working a better quality seam. The Osborne-Wallsend (Mount Keira) mine, known today as Kemira Colliery, is Australia's longest operating underground coal mine. Osborne's timing was fortuitous. Coal prices were high as a result of the gold rushes, and the miners' strike in Newcastle in 1861 sent the price of Illawarra coal from 12 to 40 shillings per ton.

Pit pony at Coal Cliff Colliery, which closed in 1991 after 114 years of operation. The Coal Cliff deposits were discovered by survivors from the shipwrecked 'Sydney Cove' in 1797, but full-scale mining of them began only in 1877, after roads and shiploading facilities were provided.

Other people followed Osborne's lead, and more mines were opened. One local businessman, Thomas Hale, actively sought out interstate and export markets, sending samples of coal to intercolonial ports and later to China. Work began in 1861 on enlarging and improving Wollongong's harbour and by 1862 four mines were operating. In 1876 Patrick Lahiff built the colony's first coke ovens. The township of Clifton came into being a year later, when Coal Cliff Colliery was developed.

Ten coal mines were working in the Illawarra by the 1880s. During that decade, mine owners had a jetty built at Red Point (now Port

Illawarra coal miners, about 1920.

Kembla), with a rail link to Mount Kembla mine. By 1890, Wollongong was the third busiest port in the colonies. The rugged country delayed construction of the railway between Sydney and the Illawarra, but it was finally completed in 1893, and brickworks and coke-making facilities were set up at many points along it.

If we need to be reminded of just how much perceptions and times change, we may consider an early description of the coke ovens: 'In their heyday it was a wonderful sight for train travellers at night to see the batteries of ovens from station to station belching their flames into the darkness'. In 1990, at Port Kembla alone $40,000,000 was being spent to improve the environmental performance of the coke ovens, a fact that reflected a change in attitudes throughout the world.

A convertor at Electrolytic Refining & Smelting (ER&S), Port Kembla.

Early Development of Secondary Industry

The Illawarra's rich endowment of coal made other developments possible. The region's industrial history began with the setting up of the Electrolytic Refining & Smelting Company of Australia Ltd (ER&S) plant in 1908. It was the largest of its kind in the British Empire, and though business began slowly, it became the nucleus of Australia's copper manufacturing industry in the First World War. Its large direct current generating plant also supplied electricity to the refinery, to employees' homes and street lights. The Company is now known as Southern Copper Pty Ltd.

A decade later, Metal Manufactures opened up in the district, fabricating copper and brass wire, telephone cables, tubes, sheets and alloys. It expanded rapidly and by 1925 its work force had grown to 800. Then came setbacks: copper prices collapsed, known Australian sources were becoming exhausted, and in 1928 Mount Lyell opened its own copper smelter at Queenstown, in Tasmania.

By the early 1920s, Wollongong's population had reached 6,700. The coal industry continued to dominate the region, with a third of the work force. The industry fell on hard times in the 1920s, as it did in the Hunter Valley. Export trade fell away after the First World War, the cost of Australian coal having trebled between 1915 and 1920. Fortunately, the decline in the coal industry was offset by the expansion of secondary industry.

A Turning Point: Steel

In the early 1920s Hoskins Iron & Steel Co. Ltd. decided to relocate its steelworks from Lithgow to a seaport. The company had purchased the Wongawilli Colliery in 1916 and already looked to the Illawarra for its coke. It chose Port Kembla, then a tiny cluster of houses, as a suitable site for an integrated steelworks.

Unable to fund the steelworks by itself, Hoskins formed Australian Iron & Steel Ltd (AI&S), with two British iron and steel companies and the Australian firm Howard Smith Ltd.

In contrast to the steelworks established at Newcastle by BHP, where the American David Baker incorporated the latest developments from

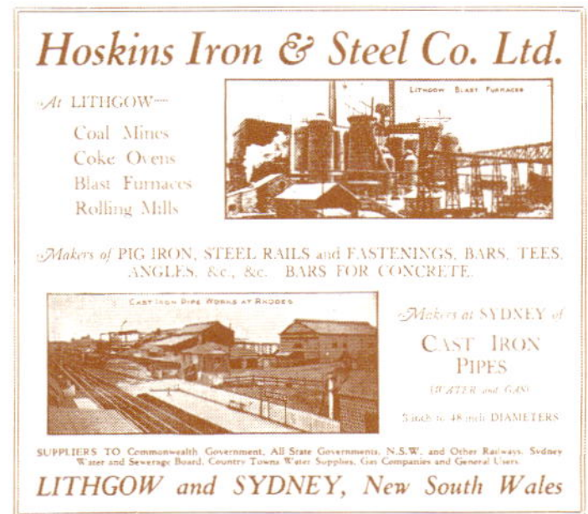

Advertisement for Hoskins Iron & Steel Co. Ltd, Lithgow. Hoskins moved to Port Kembla in the 1920s, and formed Australian Iron & Steel Ltd in association with British interests.

Early days at Port Kembla steelworks. No. 1 blast furnace and blower station on the right, spun pipe plant on the far left, rolling mills under construction in the centre.

the United States, plant for the Port Kembla works was ordered from England and laid out on the lines of a British steelworks. With Hoskins' British partners, this was not surprising, but it proved a disadvantage.

Ironically, though demand for Illawarra coal had been boosted by the application of steam power in industry, precious little local coal was used in building the new steelworks. 'It was a common sight at Port Kembla to see horses and drays and picks and shovels doing the work; the company had one steam shovel and one steam jib crane' one observer noted.

Australian Iron & Steel Limited was launched with a flourish in May 1928 when a lavish prospectus was issued. The first consignment of iron ore from South Australia arrived and the blast furnace was commissioned in August. With its 800 tons per day rating, the blast furnace was the largest and best equipped in the British Empire. Unfortunately for Hoskins, the Depression hit Australia in 1929. The first steel was made at Port Kembla in 1931, and the company struggled in the very difficult economic climate. Four years later BHP acquired all AI&S shares, and quickly set about expanding the works.

The Illawarra population grew rapidly with the growth in secondary industry. Between 1921 and 1933 it increased by one third, to almost 43,000. The increase was more dramatic in Wollongong and Port Kembla,

Port Kembla steelworks at war. Forging a gun barrel on 2000-ton press.

Junction of two electric haulage lines in an undergound coal mine at Bulli.

which recorded population increases of almost 70 per cent and 60 per cent respectively. And in spite of AI&S's problems, tonnages of exports from secondary industries exceeded coal shipments from Port Kembla in 1934 for the first time.

Expansion of the steel industry in the late 1930s saved some of the coal mining villages from virtual extinction. Associated enterprises such as British Tube Mills Ltd, Stewarts and Lloyds Ltd, and John Lysaght Ltd established themselves in the area. Commonwealth Oil Refineries Ltd (COR) built a petrol blending plant in 1937, utilising the benzol produced by the coke ovens. Clearly though, the district's economy was dominated by BHP, as the major employer and coal producer.

On the eve of the Second World War, Australia's steel industry had reached an all-time high level of prosperity. Output from Port Kembla stood at 600,000 ingot tons, and some departments were better equipped than their equivalents at Newcastle. The war tested the ingenuity of those at Port Kembla, and saw many products made for the first time.

No. 4 blast furnace, Port Kembla steelworks 1954.

Developments in Coal

Between 1907 and 1938, annual coal production in the region averaged 2,000,000 tons. During the war years this increased dramatically, the steelworks requirements alone leaping almost 40 per cent from 340,000 tons in 1935 to 475,000 in 1946.

Expansion and mechanisation of the collieries during the war required much new equipment. As this could not be brought from overseas, partly because it was needed there, partly because of shipping problems, the steel industry quickly arranged for Australian manufacturers to fill the gap.

Teeming molten steel into ingot mould at Port Kembla steelworks, 1955, using 150-ton capacity ladle at No. 1 open hearth.

The manufacturers had of necessity used existing mining equipment as models, and the most suitable proved to be those of the Jeffrey Manufacturing Company. Early in 1947, officers of BHP met with Jeffrey people while on a visit to the United States, and their candid exchange of information was impressive. According to the *BHP Review*:

> The Jeffrey people, fully understanding the situation, and with characteristic large mindedness, made available details of

improvements they had subsequently developed in their machines. Our representatives were able to reciprocate with examples of modifications resulting from their Australian experience.

Mine cars also had to be built. Production of these was shifted from Burwood Colliery to AI&S where cars of up to 10 tons capacity were made. The largest mine cars then in use in Australia, the first of them went into operation in Wongawilli Colliery. The first diesel locomotive in an Australian coal mine began work at Mount Keira in 1942. It too was built locally, by Malcolm Moore Ltd. Metal Manufactures Ltd supplied tinned copper wire core for field telephone wire during the war.

Post-war Developments

The Port Kembla works were developed further after the Second World War. Essington Lewis, CH, had long dreamed of a tinplate industry for Australia, and this began production in 1957, on reclaimed swampland, making Port Kembla the country's largest steelmaking centre. One development followed another: the hot strip mill, tinplate production, electrolytic tinning, Basic Oxygen Steelmaking (BOS) and continuous slab casting.

The downturn in the steel industry's fortunes in the early 1980s brought hardship to the Illawarra. This was inevitable, given the region's reliance upon heavy industry at the time. But new opportunities came with rationalisation of the steel industry, and new businesses started up.

The steel industry's recovery was strong. A modernisation programme at the Port Kembla works, of the order of $1 billion, was completed, with world class technology installed. Port Kembla remained Australia's major steelmaking centre. Annual raw steel capacity stood at 3,800,000 tonnes, and it was the nation's supplier of flat products, including plate, sheet steel, feed coil, and tinplate.

Hot metal charging operation at the basic oxygen steelmaking (BOS) plant, Port Kembla.

Associated Developments in the Coal Industry

The Illawarra's coal underpinned all these developments. Over the years, it had seen some remarkable advances in mining techniques, practices and productivity, some of them hastened by bitter industrial disputes.

Mining began as a pick and shovel operation. Coal cutters and mechanical loaders came in the 1940s, continuous mining in the 1950s, longwall mining in the 1960s. The advent of roof-bolting enabled greatly-improved ventilation systems. In the 15 years from 1963 to 1978, output per man per shift increased from five to eight tonnes.

By 1990 Illawarra coal was won by a combination of longwall and continuous mining techniques. Raw coal production from the southern coal fields was 16,600,000 tonnes in 1988-89, representing approximately 20 per cent of total production in New South Wales. Exports of coal from Port Kembla were 8,800,000 tonnes in that year. The 1980s saw dramatic increases in productivity with output increasing to more than 13 tonnes per man per shift.

Future Outlook

The Illawarra's mineral heritage will continue to exert a powerful influence on its fortunes. In 1990, the New South Wales Coal Association was cautiously optimistic about the outlook for the coal industry.

BHP's capital expenditure programme for the Port Kembla works involved bringing annual capacity up to 4,500,000 tonnes, as well as upgrading parts of the works. Although lower domestic demand posed challenges, the works were once again internationally competitive and were expected to continue to perform well. The rich natural endowments of the region will continue to contribute to its well-being.

I am indebted to Vicki Lukins and Les Jones of BHP Steel for their invaluable assistance in the preparation of this paper, which was presented in Wollongong on 17 September 1990.

Modern miners prepare to take a half-hour ride to the coal face at Coal Cliff Colliery.

Cobar Copper: Mines in the Outback

More than a decade before Charles Rasp discovered the rich silver-lead deposits of the broken, ragged hill in the Barrier Ranges which became the world-famous mines of Broken Hill, Thomas Hartman and Charles Campbell had found Cobar's wealth in copper and it was making a significant contribution to Australia's prosperity.

Hartman and Campbell were Danes who ventured to Australia to seek their fortunes on the goldfields, failed to do so, and became well sinkers in inland New South Wales.

The well sinkers noticed unusual blue and green colours beside one waterhole, and took samples. As they continued south they met Mrs Kruge, who had been a 'bal gal' (a girl surface worker) at Cornish mines. She recognised the specimens as copper ore and they sent samples to Adelaide for assay.

Great Cobar smelter in 1912.

Copper was confirmed and a new mining field, almost in the geographical centre of New South Wales, was established. The Aboriginal name Kubbur became Cobar. The Cobar Copper Mining Company sent ore to Louth or Bourke to go by river boats along the Darling and Murray Rivers to the smelters in South Australia.

Many prospectors came to the area, resulting in the discovery, in late 1870 or early 1871, of the United (Occidental) and the CSA mines, the latter named for the Cornish, Scottish and Australian origins of the initial proprietors: Henry and Richard Nancarrow, George Gibb, and John Connelly. High transport costs to the river and then to the smelters, and delays when the rivers were too low, led to the setting up of smelters on the mining field.

By 1877 Cobar had 500 people. As activity expanded, this grew to about 3,000 in only six years. The village had become a town, and nearly 900 people worked at the Great Cobar mine.

Charging furnaces with coke at Great Cobar in 1912.

Annual production of blister copper was about 2,400 tons from the numerous smelters. The voracious appetites of the smelters and boilers for wood required private narrow-gauge wood carting railways that extended up to 15 kilometres from the smelters. By 1875 Cobar axemen were chopping 50,000 tons of firewood a year. As the trees disappeared, the combination of a growing rabbit population and the summer winds depleted the topsoil of the plains.

Some wealthy men became even wealthier as the Cobar field developed. The river boats passed through towns along the Darling River, notably Wilcannia. Money was made, and no doubt lost, by buying shares on the hotel rumours of men on the boats.

Tapping the copper smelter at Great Cobar.

The electric locomotive used to haul and tip the slag ladles at Great Cobar early in the century is now on display at the Cobar Museum.

Copper Bonus

The New South Wales Arbitration Court awarded a copper bonus to the miners of Cobar in the years of high copper prices, 1906-07. It was similar to that paid on the Wallaroo-Moonta field in South Australia, and based on Cornish copper production practices. The idea spread to Broken Hill, where a lead bonus was first paid in 1925.

The Cobar district copper-gold mines remained in production during the First World War. Locals who joined the Army and served overseas received letters stating that they would have jobs when they returned. But the big mine, Great Cobar, closed in March 1919 even as some returning servicemen came home. By 1921, all the Cobar mines had closed.

The smelters at the CSA mine treated ore from mines in the district until one Saturday afternoon in March 1920. No-one was underground when a fire began, perhaps spontaneously in the combustible ore. The mine was forced to close, and the smelters were sold as scrap to the BHP steelworks at Newcastle. The town declined, with residents leaving, hotels and shops closing, but it survived as a service centre for the surrounding pastoral properties.

Resurgence in the 1930s

Higher prices for gold in the 1930s brought miners, promoters, speculators and others to old and new goldfields. Cobar and district came to life again with gold.

In December 1933, New Occidental Gold Mines N.L. was formed to acquire leases from Occidental Gold Mines, later from New Cobar Gold

Mines and then the Chesney and Young Australia leases. It was able to pay modest dividends, even with gold at the low price it then was. By 1938 three mines had re-opened to work gold and copper. Copper was required during the Second World War and the mines remained in production. But by 1952 they had again closed.

Two well known members of The Australasian Institute of Mining and Metallurgy worked there in the post-war years. They were Emeritus Professor John Morgan who was President of The Institute in 1981 and who presented the 1980 President's Award to Jim Woodcock, at one time Research Metallurgist at New Occidental.

Relic of a ten head stamp battery used to crush gold ore near Cobar.

Cobar's Third Life

Cobar's third life began in the 1950s, when Broken Hill interests saw new opportunities there.

In the 1950s many believed that the mines of Broken Hill had a limited life, especially Broken Hill South Ltd (BHS). In 1956 the company appointed its first geologist, Basil Lewis, formerly with Western Mining Corporation Ltd. He was given the task of extending the life of the mine by exploring initially within the mine leases, then in the Broken Hill area, and later anywhere in Australia. His investigations included Cobar.

In a way, mining history had gone into reverse. In 1885, Captain W. H. Morish had left Cobar to become the first manager of the Broken Hill South Silver Mining Company Ltd, which later became Broken Hill South Ltd. Cobar miners had once gone to Broken Hill to deepen, under contract, BHP's Rasp Shaft.

Mount Pleasant gold mine, Cobar.

CSA mine, treatment plant and service facilities at Cobar.

Now Broken Hill based men were to help Cobar. In 1956, Broken Hill South Ltd acquired options over leases and freeholds from New Occidental Gold Mines. Lewis reinterpreted the diamond drill results from exploration of the old CSA leases by a Consolidated Zinc subsidiary and further exploration began. The programme was restricted because of the low metal prices in the late 1950s.

By 1961 diamond drilling had indicated more than 18,000,000 tonnes of better than three per cent copper ore, extending to a depth of at least 1,000 metres. A decision to develop and mine the CSA orebodies was made in October 1961 by Cobar Mines Pty Ltd, then owned by Broken Hill South Ltd (76.66 per cent) and Consolidated Zinc Ltd (23.33 per cent). Shaft sinking began in February 1962 and by the end of May 1965 the concentrator was operating. The CSA mine was again in production and became the lifeblood of Cobar and district.

No. 4 level at Mount Boppy gold mine, Cobar.

New Technologies

Cobar's population grew from 2,200 to 4,000, and about 500 new homes were built. Most mine personnel were housed in the town, some 11 kilometres from the mine. Cobar was again a prosperous community. Instead of a copper bonus, a quarterly cash prosperity bonus was introduced by Cobar Mines. Water for the metallurgical plant was brought from the Macquarie River, giving the town too a reliable water supply.

Technology new to Australian mining and metallurgy was introduced to Cobar's CSA operation. A mechanised, trackless method of cut and fill mining with inclined access to the working areas was selected, the inclines replacing conventional vertical and horizontal development. A large circular, concrete-lined shaft enabled the diesel equipment to be taken underground. This type of shaft and underground machines that did not run on rails were innovations to mining in Australia.

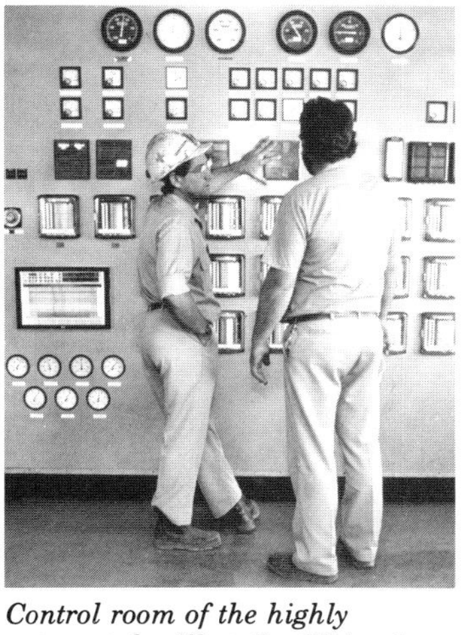

Control room of the highly automated mill at the CSA mine Cobar in 1990.

New equipment usually brings problems, and Cobar had its share. The mine became a research laboratory, but the researchers wore hard hats rather than white coats, and worked underground. The many improvements they made included underground workshops for repairs and maintenance, to keep lost working time to a minimum. The concentrator was equipped for autogenous grinding, with extensive instrumentation throughout the plant.

The 'researchers' introduced automatic controls wherever feasible, with underground crushing and the grinding and flotation plant on the surface monitored by television. The friction hoisting system for ore used rope guides, with single point suspension of the electric cables hanging down the shaft. Such approaches were commonplace by the 1990s, but they were highly innovative in the 1960s.

The autogenous mills were also a first in Australia. They were sited outdoors with substantial savings in capital costs, better access for maintenance, and reduced noise.

Spreading out from Cobar

In 1972 an aeromagnetic survey over exploration licences to the north-west of Cobar's CSA mine located 25 magnetic anomalies. Ground magnetic surveys and auger drilling in 1973 revealed the presence of anomalous lead values in the bedrock. Following diamond drilling, the EZ Company reported a resource of 27,000,000 tonnes containing 5.6 per cent lead, 8.6 per cent zinc, and 135 grams of silver to the tonne. The company decided in July 1980 to develop a mine at Elura, about 40 kilometres north-west of Cobar. It increased its production steadily in the 1980s.

The Peak gold and base metal field was again to become a producer with the CRA announcement in June 1990 that a mine was to be developed to produce annually about 100,000 ounces of gold in bullion and concentrates, 2,500 tonnes of copper, 4,800 tonnes of zinc and 3,600 tonnes of lead in concentrates.

Peak gold mine near Cobar, during development stage. CRA sank an exploration shaft 740 metres before it decided to go ahead with the mine.

Cobar people felt confident that the third life of the district be extended by these developments. And who knew what other economic deposits remained to be found in this highly mineralised region? The odds were that the Cobar district would remain on the mineral map of Australia for a long time.

History Recorded

As the mines and the town of Cobar and the pastoral industries developed, there were those who recorded events in words and photographed the happenings. I am especially pleased to acknowledge William Clelland and his fine volume called *Cobar: Founding Fathers.* He is a member of The Institute and worked on the CSA mine.

In his preface he says that he was stimulated by an unfulfilled promise to Ian Solomon, who died in 1979. We now have a fine volume of the history of the area until the end of the 19th century. The author leaves a challenge for writers and historians – and Institute members in general – to continue the story throughout the 20th century.

I am indebted to John Dew and Gilbert Ralph, MBE, for their invaluable assistance in the preparation of this paper, which was presented at Cobar on 17 July 1990.

Broken Hill's Great Contribution

In 1844, Captain Charles Sturt led the first expedition of Europeans to the far west of New South Wales. He saw a prominent, rugged hill topped with dark coloured rock, which he recorded in his journal as a 'broken hill'. He collected mineral specimens, but was later forced to jettison them along with other non-essentials, such was the hardship of his journey in the searing heat of a waterless summer.

The large Mount Gipps sheep run, taken up in 1867, encompassed the 'broken hill'. In that same year a few hardy gold prospectors climbed it, only to turn away, disappointed. Other prospectors found silver 15 years later some 30 kilometres to the west and by 1883 Silverton was a bustling frontier town. To the silver miners the 'broken hill' was just a hill of mullock, worthless rock. But on 5 September 1883 a Mount Gipps Station hand, Charles Rasp, collected pieces of the dark rock, noticed their unusual weight, and concluded that the mineral was the black oxide of tin. Next day he returned and pegged a mineral lease.

Captain Charles Sturt. On his 1844 exploration of western New South Wales he recorded in his journal a 'broken hill'. He collected mineral specimens there, but had to abandon them later in his journey.

Rasp's real identity, according to recent research, was Hieronymous Salvator Lopez von Pereira, a Portuguese aristocrat and a former Prussian army officer. Apparently he had deserted, taken the name of a friend killed in action in 1870, and fled to Australia. Death was the penalty for desertion, but Rasp died peacefully in Adelaide in 1907, a rich man, his secret intact.

Six men from the Mount Gipps Station joined Rasp to form a 'Syndicate of Seven' and pegged most of the large outcrop. A small shaft exposed lead minerals with minor silver but nothing exciting.

In January 1885, rich chloride of silver was found. In the ensuing excitement, on 20 June, the first prospectus of The Broken Hill Proprietary Company Limited (BHP) was issued. The Company was floated on 10 August 1885, with William Jamieson, a government surveyor, as its first manager.

Other prospecting groups had by then pegged leases over the rest of the outcrop, and beyond. Two of the many new companies formed were the predecessors of North Broken Hill Ltd and Broken Hill South Ltd, both floated in 1885. A later starter was The Zinc Corporation Ltd, formed in 1905 to purchase and treat zinc-rich tailings dumps.

These four companies – BHP, North, South, Zinc Corporation – were destined not merely to be highly successful at Broken Hill, but also to devote their profits and expertise to new enterprises in many parts of Australia. They developed other mining fields, constructed railway lines, built smelters, launched large forestry projects and made paper.

Charles Rasp, the boundary rider who discovered the wealth of Broken Hill.

Industrial Relations

Broken Hill was a trying place to live and work. Only hardy people stayed. The bosses were hard, too. Even so, early industrial relations

BHP mine and treatment plant at Broken Hill, 1887.

were calm, with a union formed in Silverton in 1884 becoming, in 1886, the Barrier Branch of the Amalgamated Miners Association, the AMA.

Unionism grew in a working environment of poor health and safety conditions and tough management. Its growth was helped by Marxist ideas ('Workers of the World Unite') brought to Broken Hill by English labour leaders and visiting orators such as Tom Mann.

William Jamieson, first General Manager of BHP. An Aberdeen man who had been a government surveyor in the Barrier district, he resigned as manager in favour of a man with more mining experience, but he sat on the BHP board for 20 years.

The biggest of all Broken Hill's strikes began in 1919, lasted 20 months, and affected every mine. Although the main claims were about wages and hours, very real health and safety issues festered beneath the surface and during the strike a Commission of Enquiry into workers' health was appointed. The strike's end in 1920 brought a 35-hour week for underground workers, with blasting only at the end of shifts, a break of one hour between shifts to allow dust and fumes to be exhausted from the workings, and compensation for disabled and injured workmen.

From 1920, industrial relations generally improved, following the initiatives of people like Andrew Fairweather, Manager of the South Mine, whose report on *What is Wrong with Broken Hill?* led to some innovative reforms. Inefficient work practices, introduced in times of high metal prices, however, remained until much later, when the need to improve productivity and reduce costs was better understood, and agreed by all parties.

The Companies

Of the companies formed to work the great lode, some closed their mines as the ore was depleted. Others merged in an effort to survive. In 1989, the two remaining large mines were amalgamated into a new company, Pasminco Ltd. The only other company then operating in

Broken Hill was Minerals, Mining and Metallurgy Ltd, which was recovering remnant ore, mainly by open cut methods, from the Broken Hill South Ltd leases it purchased in 1972.

Three of Australia's greatest companies had their origins in Broken Hill: BHP, CRA, and North Broken Hill Peko.

BHP: The Broken Hill Proprietary Company Limited

The Broken Hill Proprietary Company Limited, soon to be widely known as BHP and to become a major force in Australia's industrial development, was the first company to operate at Broken Hill. In the first few years its shareholders were rewarded with huge dividends, while some of the profits were invested in development.

With increasing depth the silver content fell, and mining became more difficult. The Board of Directors, acknowledging a lack of expertise, but sensing the size of the problems, recruited a highly reputable, experienced mine manager, William H. Patton, from the famous Comstock mine in the United States.

As the workings went deeper, serious metallurgical problems arose, as the easy-to-smelt oxidised ore gave way to sulphide ore. It proved possible to separate by gravity methods most of the silver-lead mineral, but the zinc mineral was left in the residue ('tailings'), huge dumps of which soon accumulated on the surface.

A reward awaited anyone who could find a way to separate the zinc. In great secrecy and with great urgency, research was begun by all the companies and by some outsiders. BHP was to the fore. By this time, Guillaume Delprat was General Manager, reputedly the highest paid person in Australia. Delprat conceived the idea of flotation and Leslie

Guillaume Delprat, CBE, who became BHP General Manager in 1899. He played a major role in the introduction of flotation as a way of recovering zinc from the complex ores, and in BHP's move from mining at Broken Hill to steelmaking at Newcastle.

The first 50 tons of zinc concentrates produced at Broken Hill by flotation. Guillame Delprat, CBE, who encouraged development of the flotation process, is seated in front.

An electric locomotive hauling trucks of ore undergound at the Proprietory mine.

Bradford, a young graduate from the South Australian School of Mines, was given the task of putting the idea to practical use. By 1902, in a world first, the flotation process was being used to separate the zinc mineral from the rest of the surface dumps. The process caused it to float to the surface of the flotation chambers, enabling it to be skimmed off. By 1913 another achievement of world significance was the development of differential flotation, with the silver-lead mineral selectively floated first, then the zinc mineral. Again BHP, with Bradford, had a leading role.

The events that were to take BHP into the iron and steel industry began early in the company's development.

By 1889 it was clear that smelting could be done more economically at Port Pirie, in South Australia, than at Broken Hill. In 1893 BHP bought the Port Pirie plant of the British Broken Hill Company and by 1897 it had phased out smelting in Broken Hill.

To supply the iron oxide the Port Pirie smelter needed as flux, BHP in 1890 leased iron deposits in the Middleback Ranges, across Spencer Gulf. This development led BHP, under Delprat's leadership, to investigate steel making, using the services of David Baker, an American consulting engineer highly experienced in steel making. After spending some time in Australia, Baker recommended Newcastle as the site for the steelworks, because of its port and the proximity of good coking coal, and steel making began there in 1915.

With production from the Big Mine sporadic after the long strike of 1919-20, BHP in 1925 sold its interest in the Port Pirie smelter to the other Broken Hill companies and in 1939 it surrendered its mining leases in Broken Hill. The transition from silver to steel was complete and the wealth from the great Broken Hill lode had been used to good effect.

Minerals separation equipment at the Zinc Corporation plant, Broken Hill, 1915. Plants such as this enabled the recovery of millions of pounds worth of zinc that had been locked up in the tailings dumps from the silver-lead mines.

North: North Broken Hill Peko Ltd

The original 'North' company was formed in 1885 to develop the northern leases at Broken Hill. The name was changed several times, but it became North Broken Hill Ltd in 1912.

Conservative management was to characterise the Company. Shareholders were rewarded with good dividends while investments were made in lead smelting at Port Pirie, in helping to establish a major zinc metal producer (the Electrolytic Zinc Company, in Tasmania), and the important paper-making company (Associated Pulp and Paper Mills Ltd – APPM), also in Tasmania.

In the 1980s, North Broken Hill Ltd absorbed both Electrolytic Zinc and APPM, prior to its merger with Peko Wallsend Ltd in 1988. It became a major shareholder in Pasminco Ltd, which took over lead-silver-zinc mines at Broken Hill, Elura (NSW) and Rosebery (Tasmania), and smelters and refineries at Port Pirie, Cockle Creek (NSW) and Risdon (Tasmania). It also has interests overseas.

CRA: Conzinc Riotinto of Australia Ltd

Herbert C. Hoover, later to be President of the United States of America, visited Broken Hill as a 31-year-old engineer in 1905 and arranged the purchase of large tailings dumps containing zinc. Later that year, The Zinc Corporation Ltd was formed in London to extract the zinc from these dumps. The company was to learn that the practical application of new processes can be fraught with difficulties. Not until the fourth process tried, after five years of technical battle and financial ordeal, did it find one that worked.

The founders of the Zinc Corporation included a number of influential Australians. One of them, W. S. Robinson, a former finance editor of the Melbourne *Age* who had become a mining entrepreneur, was appointed Managing Director in 1926.

The company delayed its first dividend until 1912, having used earlier cash flow to acquire mining leases near its zinc treatment plant. Those

Amalgamated Zinc Company (de Bavay's) plant, one of the pioneer flotation plants of Broken Hill. It produced an enormous tonnage of zinc concentrates from the tailings of some of the mines.

Lead bullion ready for shipment from the BHP mine at Broken Hill. The company soon moved its smelting operations to Port Pirie, in South Australia.

leases proved to contain a huge tonnage of high grade sulphide ore. In 1936 a new company, New Broken Hill Consolidated Ltd, was formed under W. S. Robinson, to explore and, as it turned out, to mine the contiguous mining leases to the south. The new mine was operated in close association with the Zinc Corporation.

In 1949, the Zinc Corporation merged with the Imperial Smelting Corporation Ltd, of the United Kingdom, to form Consolidated Zinc, which later merged with Rio Tinto to become Rio Tinto Zinc Corporation Ltd. Its Australian associate, Conzinc Riotinto of Australia Limited, became CRA Limited, and expanded its mining and minerals interests to become one of Australia's largest companies.

W. S. Robinson, a former financial journalist who became prominent in mining and other developments in Australia.

Technology for the World

Broken Hill has contributed significant technological advances to the mining and minerals industries of the world, the most important being flotation and hydraulic fill.

Flotation came to be used in thousands of ore treatment plants throughout the world. The process has made many a metalliferous deposit economically workable and enabled production at competitive cost of such metals as copper, lead, zinc, nickel, silver, gold and others.

Not so spectacular, but of great importance to the efficiency and safety of underground mining, was the technique of hydraulic fill. It, too, was developed in Broken Hill, by Broken Hill South Ltd, after an early start at Mount Lyell in Tasmania, and came to be used in many mines throughout the world.

Many other advances have also been made at Broken Hill in conceptual geology, exploration drilling, mining, mineral processing, and environmental care and rehabilitation. Treatment of the concentrates from Broken Hill containing lead, silver, zinc and sulphur have also been the subject of innovative techniques: continuous lead refining and jarosite iron removal in zinc metal production, for example, were world firsts. The introduction of up-draft sintering enabled the manufacture of sulphuric acid from sinter gases.

During the late 1950s Broken Hill was involved in technological research with CSIRO's Alan (later Sir Alan) Walsh which gave the world atomic absorption spectroscopy. It also shared in research associated with the use of x-ray fluorescence using radio-isotopes.

Part of the zinc-rich tailings heaps at Broken Hill which for many years eluded attempts to extract the zinc.

Contributions to Australia's Industrialisation

Professor Geoffrey Blainey, AO, in *The Rise of Broken Hill*, described Broken Hill's contribution to Australia's industrialisation from 1914 to 1939, in four separate ways.

First, the profits from the rich ore body became a source of capital for other industries in Australia, largely because the companies were mostly directed from Australia. The Zinc Corporation Ltd, although directed from London, was greatly influenced by its Australian-born Chief Executive, W. S. Robinson.

Broken Hill South mine and treatment plant about 1908.

Froth and bubble of the kind that enabled metallurgists to find ways to separate zinc from the enormous tailings heaps at Broken Hill. Australian methods of flotation to separate minerals came into use all over the world.

Second, economic sites for lead and zinc smelters were far from Broken Hill and the companies were forced to broaden their activities geographically. Smelters required ironstone and limestone fluxes, and coke, so it was natural for the companies to seek good deposits of such raw materials. This led to the development of steelworks, coal mines and cement manufacturing.

Third, smelters treating sulphide ore produced as a by-product large quantities of sulphur dioxide gas, which were put to use in making sulphuric acid, a raw material for many secondary industries, including fertilisers.

Fourth, the metals from Broken Hill offered scope for further processing. Lead was turned into white lead for paint, for example, and into lead-sheathed cables and alloys.

A Place to Live

From a few huts and tents in 1885 Broken Hill grew to a population of 6,000 by 1888, the year the first train arrived. Three years later the population passed 20,000. In 1907, Broken Hill was proclaimed a city, when its population was nearly 30,000, second in New South Wales only to Sydney, and sixth largest in Australia.

In this area of recurring droughts, water was always scarce. In the early years Broken Hill water was often contaminated by sewage, and typhoid infected those who drank it unboiled. Not until the 1950s was a reliable water supply provided by the construction of a 100-kilometre pipeline from the Darling River.

Run-off from roofs was undrinkable because it contained lead from dust blown from the smelters. Lead poisoning affected mine and smelter workers in the early years and it was not uncommon for a man to throw a 'lead' fit in the main street.

Undergound party in the South Mine at Broken Hill, 1919. The ladies were specially privileged, for women were not welcomed underground. The presence of the mine manager, W. E. Wainwright (with lady holding his arm), no doubt helped these ladies have the custom suspended.

Concentrator plant at the Zinc Corporation mine.

With wood in demand for household cooking, heating, steam generation, and underground support in the mines, bare rock and sand soon surrounded the town. Once all the mulga was gone, even the roots of salt bush were dug up for household fuel. New shoots of vegetation were eaten by the goats, sheep, cows and horses. The slightest wind blew fine dust through every crack in the corrugated iron houses. Sporadically, there were severe 'black-out' dust storms.

To overcome the dust problem, A. J. Keast, CBE, General Manager at the Zinc Corporation, under the strong influence of Maurice (later Sir Maurice) Mawby, CBE, in 1936 backed the concept of Albert Morris, Chief Assayer at the Central Mine and an amateur botanist, to establish a regeneration area around the town and the mine leases. The City Council co-operated and an area around the city was fenced against rabbits and stock. Seeds in the sandy soil germinated after rain and new shoots survived despite poor rainfall in the early years, making the regeneration area a huge environmental success.

The women of Broken Hill supported their husbands' industrial struggles, bore the hardships, and raised their families. Career opportunities for young women were few: nursing at the Broken Hill and District Hospital, secretarial work, and teaching (after training in Sydney). Choral societies encouraged young singers like June Gough, whose career was launched from Broken Hill, a fact she acknowledged in her stage name, June Bronhill.

The city became a pleasant and healthy place, the service centre for the whole West Darling District providing hospital, educational, administration and recreational facilities, with good communications to Adelaide, Melbourne and Sydney.

Broken Hill and The Institute

In 1893, the mine managers of Silverton and Broken Hill led the founding of what later became The Australasian Institute of Mining and Metallurgy, their objective being to raise professional and ethical standards in the minerals industry. Many leaders of The Institute and the mineral industry were to come from Broken Hill.

Broken Hill has an important place in Australia's mineral heritage and its contribution to Australia has been tremendous. It will continue to serve Australia well, and future generations will continue to recognise that.

Stamp issued for the 75th anniversary of Broken Hill, 1958.

I am indebted to Don Fairweather, former Manager of Broken Hill South Ltd. and Gilbert Ralph, MBE, for their invaluable assistance in the preparation of this paper, which was presented in Broken Hill on 18 July 1990.

A raise drill in use at the New Broken Hill Consolidated mine in 1974. These hydraulically operated drills are now in common use to drill vertical holes between levels to act as ore passes, ventilation shafts or access ways.

Australian Capital Territory

Canberra welcomes a new mine.
Prime Minister the Rt. Hon. Sir Robert Menzies, KT, CH,
at the opening of Mary Kathleen uranium mine,
Queensland, 1958.

Government and the Industry

Not so long ago, 'Governments' to the mineral industry meant 'State Governments'. The States own the minerals in the ground. They control who can explore for minerals, where, and under what conditions. They grant rights to mine. They charge licence fees for prospecting and mining and levy royalties on production.

An early clash between miners and government. Eureka Stockade, the gold miners' armed insurrection against the Victorian Government, Ballarat, 1854.

It was the way Victoria used its powers that led to the gold miners' armed insurrection at Eureka Stockade, on the Ballarat goldfield, in 1854. The miners revolted against discriminatory taxation. In more recent times the Federal Government has used its control over exports, and various other devices, to veto or restrict some mineral developments. These Federal actions have in effect over-ridden State rights in some areas.

Government Departments

The discovery of minerals, starting with coal in 1797 and lead ore in 1841, prompted the Colonial Governments to establish Geological Surveys, the earliest being in Victoria in 1852. The Colony of South Australia had, through the South Australian Company, already appointed in 1836 a German scientist, Johann Menge, 'to explore the natural productions of the Colony above and below ground and to superintend the working of quarries of stone and slate, the boring of water, metal, lead, coal, and other mines'.

Johann Menge on one of his rambles, with geological hammer in his hand and bag of mineral samples on his back.

All Colonial Governments appointed Government Geologists: Alfred Richard Cecil Selwyn in Victoria in 1852, Charles Smith Wilkinson in New South Wales in 1875, Richard Daintree in Queensland in 1868, Gustav Thureau in Tasmania in 1882, Henry Yorke Lyell Brown in South Australia in 1882, Harry Page Woodward in Western Australia in 1888, and Dr H. I. Jensen in the Northern Territory in 1911. The first Federal appointment was that of Dr W. G. Woolnough as Geological Adviser to the Department of the Interior in 1927. He was succeeded by Dr Harold G. (later Sir Harold) Raggatt in 1941.

Another landmark Government action occurred in 1934 when the Commonwealth, Queensland, and Western Australia combined to form the Aerial, Geological and Geophysical Survey of Northern Australia, with P. B. Nye as Executive Officer. In the next six years, extensive geological and geophysical surveys of the Australian continent north of 22° were carried out. As well as the general and annual reports, over 180 individual local survey reports were published.

In 1946 the Geological Section of the Department of the Interior and the Northern Australian Survey were merged to form the Bureau of Geology, Geophysics and Mineral Resources, under the direction of Dr Harold Raggatt. The Bureau's Annual Reviews, first published in 1949, are a valuable record. Apart from his direction of the Bureau, and his later role as Secretary of the Department of National Development from 1951, Sir Harold is remembered for his book, *Mountains of Ore*. He was the recipient of the Institute Medal in 1964.

The Bureau of Geology, Geophysics and Mineral Resources became the Bureau of Mineral Resources (BMR) and together with the Geological Surveys, the latter mostly combined with State Mines Departments, undertook the systematic recording and interpretation of the geology and geophysics of regions, mapping, the investigation of geological, mineral and water resources, the publication of maps and reports on those resources, and the provision of technical advice to governments.

Henry Y. L. Brown, who in 1882 became first Government Geologist in South Australia.

Government Encouragement

For over a century Australian Governments encouraged mineral exploration and development, starting with rewards for the discovery of gold from the 1850s. The gold mining industry was encouraged by exemption from corporate income tax from 1924, an exemption which ended in 1990. In 1954 a subsidy on gold production was introduced for a time, in recognition of the price of gold having been held fixed by Government fiat, while rising production costs threatened to close the industry. In 1943 the Federal Government offered a reward for uranium discovery, and in 1957 petroleum search subsidies. In the 1960s the tax regime encouraged mineral exploration and development. The idea was simple enough: forego some early tax revenue to ensure a greater return later.

Without such exploration and development incentives, it is doubtful that the discovery and development of Bass Strait oil and gas would have taken place when it did.

Government Mining Enterprises

State Governments have a long history of direct involvement in coal mining to supply their electricity authorities. The Electricity Commission of New South Wales has operated steaming coal mines for decades. In Victoria, the State Coal Mine operated at Wonthaggi from the late 1890s to 1968, and the State Electricity Commission of Victoria opened up the vast brown coal deposits of the Latrobe Valley from 1921. In South Australia, deposits of low grade coal at Leigh Creek were brought into production by the Electricity Trust of South Australia.

During the Second World War, for strategic purposes, the King Island scheelite deposits were developed with Federal funds. After the war the Commonwealth and Tasmanian Governments for similar reasons set up the Bell Bay aluminium refinery and smelter.

Harry Page Woodward, who in 1888 became first Government Geologist in Western Australia.

Industry Associations and Government

Because minerals are the property of the States, the early industry relationships were with State Governments and the earliest industry associations had local or colonial bases.

One of the first was the Amalgamated Mining Managers' Association of Victoria, which the Broken Hill Mining Managers' Association formed in 1886. Very early in its life it became involved in negotiations over wages and other working conditions in addition to other matters of common interest. It became the oldest surviving body of its kind in the Australian mining industry.

Charles S. Wilkinson, who in 1875 became first Government Geologist in New South Wales.

The Victorian Chamber of Mines was formed in 1888. In Western Australia, a Chamber of Mines was established in Coolgardie in 1895, a similar body in Kalgoorlie in 1896. The two amalgamated in 1901 as the Western Australian Chamber of Mines, with headquarters in Kalgoorlie, where it remained until it transferred to Perth in 1970.

There was a long gap until the Queensland Chamber was formed, in 1948, followed by the Northern Territory in 1955, South Australia in 1979, New South Wales in 1981, and Tasmania in 1984. The Victorian Chamber, which became inactive when gold production in Victoria virtually ceased in the 1950s, was re-activated in 1986.

The first national representative body, the Australian Mines and Metals Association (AMMA), was formed in 1918. By the standards of the day its charter was extremely broad. It could represent the industry at Federal and State Arbitration Tribunals, deal with legislation and other measures affecting the industry, gather statistics and information, promote the interests of the industry and public discussion of it. Most significantly, it could represent the industry with Government Departments and other relevant bodies. It was envisaged that AMMA would develop into a peak industry organisation, much as the Australian Mining Industry Council (AMIC) was to become. In the event, AMMA concentrated its efforts in two areas: industrial relations and tariffs. As the industry's main revenue came from exports, and world prices were

Mines Department office staff, Darwin. Dr H. I. Jensen, appointed in 1911 as first Government Geologist in the Northern Territory, is in the centre.

Members of Parliament at an early brown coal mine in Victoria's Latrobe Valley.

something it could not influence, its profitability depended largely on containing of costs, of which labour and equipment were the major components.

Federal Government Involvement

Despite their limited involvement in the mining industry before the 1960s, Federal Governments did take some important steps.

In the late 1930s, the Federal Government banned the export of iron ore, to conserve what were regarded as Australia's limited resources of high-grade ore. As the single domestic consumer (BHP) was already self-sufficient, any newly-discovered deposits had no market once exports were prohibited. Exploration for iron ore ceased and all hope of discovering more reserves came to an end. This was the exact opposite of what Australia needed, but it took more than two decades for this simple logic to penetrate. The ban was lifted in December 1959, largely because of the efforts of Sir Harold Raggatt and several industry leaders.

Another area of Federal involvement in mining was taxation provisions that enabled exploration expenditure to be offset against income from mining and much of the taxation on new projects to be deferred.

Western Australian Chamber of Mines office, Coolgardie, about 1900.

Industry and Public Policy

In the 1960s, the industry's profile was low, so low as to be almost invisible. Early in 1962, after its near-defeat at the election in the previous year, the Menzies Government initiated a series of conferences between senior Cabinet Ministers and major industry organisations, such

Sand mining plant on Fraser Island. Queensland Titanium Mines Pty Ltd was forced to stop mining when the Fraser Federal Government banned the export of minerals from Fraser Island. The plant had to be dismantled and over 300 people retrenched.

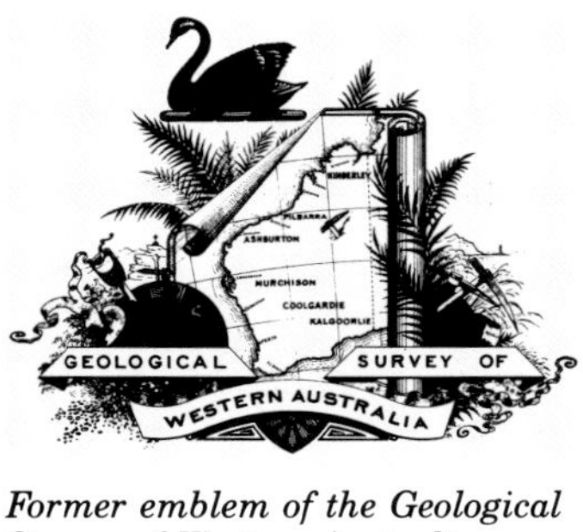

Former emblem of the Geological Survey of Western Australia.

as the Chambers of Manufactures and Chambers of Commerce, farmers, banking organisations, and the Australian Council of Trade Unions, but the mining industry was not considered important enough to be invited.

Shortly afterwards the Government opened its economic policies to public scrutiny by the Committee of Economic Inquiry, under the chairmanship of Sir James Vernon, AC, CBE. The fourteen topics under reference included the availability of resources, savings, foreign investment, costs, exports, the balance of payments, and the internal effects of protection. Again, the mining industry was not invited to make a submission and, again, at first the industry saw no reason to offer its views. Wiser counsel prevailed, however, and in 1963 the AMMA, under the Presidency of Hugh Murray from Mount Lyell, made a submission.

By then the great development of the industry in the 1960s had started: gigantic projects were either in progress or planned at Weipa, Gladstone, Bell Bay, the Darling Range, Kwinana, Point Henry, Mount Tom Price, Mount Newman, Mount Goldsworthy, Bass Strait, Kambalda, the Bowen Basin, Renison, and Mount Isa. After having a vital, but virtually unseen, role in the economy for the first six decades of the century, mining was about to leap into the centre stage of national development.

Australian Mining Industry Council

The lack of an effective industry voice was, at first, more apparent to the Government than to the industry. On accepting his Honorary Membership of The Institute in 1962 and the Medal in 1964, Sir Harold Raggatt twice called on the companies to get together. The Minister for National Development, Hon Sir David Fairbairn, KBE, DFC, later dropped strong hints that he would support such a move.

In this climate the Queensland Chamber of Mines, led by its President, Sir Clarence Byrne, OBE, DSC and supported by Sir George Fisher, CMG, convened meetings in 1966 and 1967 that led to formation of the Australian Mining Industry Council and the appointment of Paul Phillips, OBE, then a Deputy-Secretary in the Department of Trade, as Executive Director. He took up his duties on 1 July 1967.

In the second half of the 1960s, mining was rarely out of the news, with new mines, new towns, new ports, new railways, new export contracts, new minerals, new processing plants and new national wealth. Miners almost became popular heroes and mining suddenly had a high profile, too high in fact for its own good.

Prime Minister Rt Hon Sir Robert Menzies, KT, AK, CH, QC, opening the BHP hot strip mill, Port Kembla, in 1955.

Changed Attitudes

The notion took hold in the early 1970s that the minerals industry was a source of unlimited and easy to get wealth and this brought in its wake unfavourable changes in both government and public attitudes. With the Federal Government in particular becoming deeply involved in mineral industry matters, the Australian Mining Industry Council soon had its hands full. It had been formed just in time.

The Council reacted quickly to the popularising of environmental issues in the late 1960s. It appointed its first full-time environmental officer in 1969, formed the Environmental Quality Committee in 1970, and organised one of the first conferences on environmental issues in 1971.

A Technical Officer from the Department of Mines and Energy, Darwin, monitoring a waste rock stockpile with a gamma ray dosimeter.

The initial environmental attacks were directed at the mineral sands industry, no doubt because of its visibility near population and tourist centres in the coastal regions of New South Wales, Queensland and Western Australia.

Very soon, other surface mining came under attack: the bauxite mines at Weipa and the Darling Range and the coal mines in the Bowen Basin, and later all mining, but it was the sand miners who bore the brunt of the initial attack. In 1970, in an unprecedented action, a mining licence at Cooloola in Queensland was refused after the issue of an exploration licence, all conditions of which had been fulfilled, and after considerable expenditure on proving the deposit.

G. Paul Phillips, OBE, (left) Executive Director of the Australian Mining Industry Council from 1967 to 1981, with Deputy Prime Minister, Rt Hon Doug Anthony, CH.

Worse was to follow. On Fraser Island, off the Queensland coast, a major mineral sands deposit had been proven and a mining lease was granted. With only one per cent of the island to be mined over 20 years, no more than one two-thousandth part of the island would be disturbed in any one year. The usual strict environmental and rehabilitation safeguards were imposed.

Capital was raised and invested in mining facilities, and production and exports began in 1975. Then, in 1976, the project was halted, ostensibly on environmental grounds, in what was arguably the worst government decision affecting the mining industry in Australia's history. It is the only instance of a mining operation in Australia closed down by government decree.

Both the Cooloola and Fraser Island decisions were made by conservative governments. But neither side of politics has a monopoly in dishonouring their commitments. Later experience at Coronation Hill showed that both sides of politics are guilty of that.

Encouraging mineral exploration and development through the taxation system attracted criticism, paradoxically because it was successful. The Whitlam Government referred the whole subject of mining

Prime Minister, Hon R.J.L. Hawke AC, MP, (left) with Hugh M. Morgan, AO, President of the Australian Mining Industry Council from 1982 to 1984.

Coronation Hill, Northern Territory. Experience here showed that neither side of politics has a monopoly on not honouring its commitments.

taxation to the Industries Assistance Commission in mid-1974. But even before the hearings began, all the incentives were removed by the 1974 Budget.

For reasons which were never explained, the chief critics failed to appear before the Commission, whose recommendations would have gone a long way towards restoring the earlier system. Virtually no action was taken.

One of the first actions of the Whitlam Government after assuming office in December 1972 was to set up a Royal Commission into Aboriginal land rights. The Commissioner was not asked to inquire into whether land and mineral rights should be given to Aborigines, but merely into how these should be given.

Some Aboriginal communities in the Northern Territory have benefited financially from their mineral rights, but a great many more have not. The provisions do nothing for the large numbers of Aborigines who live outside the Territory, least of all for those in the cities and country towns. Even those with mineral rights are learning that these rights can be frustrated and over-ruled on environmental grounds. For the mining companies, the problems of reaching agreement with the land-owning communities have retarded development in the Territory, to the disadvantage of the whole community, including the Aborigines.

In 1971, several major discoveries of uranium in the Northern Territory were high grade, large, and near the surface, which made them amenable to open-cut mining. Anywhere else in the world, such deposits would have been immediately brought into production, as rising demand was taking the price of uranium to unprecedented heights in the mid-1970s.

But our governments dithered: the McMahon Government because it was thinking of creating a national park in the region, the Whitlam

Government because it wanted the Aboriginal land rights settled first, the Fraser Government presumably because it feared the anti-nuclear lobby. The usual Australian Government pastime of lengthy enquiries into everything allowed a decision to be avoided for many years.

The result was to miss the peak of the market. The main beneficiaries were South Africa and Canada, who willingly filled the breach, surely a classic example that time, which appears to be of no consequence in politics, is vital in commerce. The outcome was that Australia had 'good' uranium, produced from Ranger, Nabarlek, and Roxby Downs, while its 'bad' uranium stayed in the ground at half a dozen other sites around Australia.

Prime Minister, Hon E.G. Whitlam, AC, QC, opens Western Mining's Kalgoorlie nickel smelter, 1973.

Other Government actions in the early 1970s which may exercise the minds of future historians, and none of which did anything positive for the Australian community, included:

- continual criticism of foreign investment, without any attempt to set guidelines for acceptable investment
- creating of a vast network of export controls in the hope of dictating prices of minerals
- attempting to earn 'easy money' by direct participation in the mining industry through formation of the Petroleum and Minerals Authority.

Future Outlook

All miners are optimists. If they were otherwise they would not be in the business. Australia needs such optimists now as much as ever. Further growth of the minerals industry is one of the most likely contributors to overcoming our very serious current account deficits.

Governments can encourage or discourage this growth. It is to be hoped that the recent discouraging experiences will not be the pattern for the future, because Australia cannot afford it. Most importantly, the industry cannot develop and grow if large areas of potential mineral bearing land are excluded from exploration and development. Mineral deposits can only be found and mined where they occur.

Anti-uranium mining protesters at Roxby Downs, 1983.

It is one of the important responsibilities of people in the industry, including members of The Institute, to explain the vital importance of the mineral industry to the public. Sensible public policies depend on balanced and well-informed public opinion.

I have great faith in the good sense of Australians. Provided those who know the facts speak up, I believe that this good sense will eventually prevail in relation to public policies affecting the industry.

I am indebted to Mick Game, formerly of AMIC, and Gilbert Ralph, MBE, of WMC for their invaluable assistance in the preparation of this paper, which was presented in Canberra on 21 August 1990.

Queensland

Coral Sea
Weipa
Palmer River
Mount Mulligan
Hodgkinson
Chillagoe
Cairns
Atherton
Herberton
Normanton
Croydon
Gilbert River
Greenvale
Burdekin River
Townsville
Ravenswood
Charters Towers
Mount Leyshon
Bowen
Collinsville
Mount Isa
Mary Kathleen
Cloncurry
Duchess
Goonyella
Peak Downs
German Creek
Emerald
Blackwater
Rockhampton
Mount Morgan
Gladstone
Biloela
Fraser Island
Gympie
BRISBANE
Moonie
0
300 km

The Mineral Industry and the Environment

Concern about the environment and where humankind is heading is not a recent issue. Cyprian, the Bishop of Carthage, expressed these views 1,700 years ago:

> The world has grown old and lost its former vigour. Winter no longer gives rain enough to swell the seed nor summer sun enough to toast the harvest. The mountains are gutted and give less marble; the mines are exhausted and give less silver and gold. The fields lack farmers, the seas sailors, the encampments soldiers. There is no longer any justice in judgements, competence in trades, disciplines in daily life.

Immigrants arrive at Melbourne in the 1860s in search of a better way of life.

But for most of the 1,700 years, since then, survival from hunger, disease, plagues, pestilence, war, famine and the like were more urgent than concern for the environment.

The Industrial Revolution and the consequent economic development since then have not been without many adverse effects, but no-one would doubt that its positive results far outweigh the problems and risks. This is undeniable because the average lifespan in the developed world has increased from thirty-odd years to more than seventy.

In fact, technological and economic development and our resulting longer lifespan have been so successful that the foremost world problem today is population. Today's environmental concerns arise from the legitimate desire of more and more people for better food and shelter, better clothing, better consumer goods, better health care, better communications, better travel, better entertainment: in short, a better quality of life.

Diggers at Myers Creek, west of Bendigo, in 1868. Such concentrations of people had a major impact on the environment.

Affluence and the Environment

One reason our environment ranks high amongst popular concerns today is our affluence. We do not hear of an environmental movement in Ethiopia or Bangladesh. Today's average Australian lives in circumstances enjoyed only by the fortunate few as recently as 50 years ago. We can therefore afford to take an interest in such matters.

The economist Kenneth Boulding late in his working life was able to say that 'the world today is as different to the one I was born in, than that world was to Julius Caesar's'.

The genuine desire to care for the environment, and the resulting efforts to alleviate and limit the undesirable effects of human activities, are sensible and welcome. We all live in this world, albeit briefly, and nobody wants to leave our children and grandchildren an environmental legacy which is unsatisfactory and could have been avoided. The mineral industry not only agrees with this but has in many instances taken the lead in environmental care and rehabilitation of areas disturbed by its activities. In some cases, this has preceded by decades the quite recent public recognition of these issues.

Regrettably, genuine public concern has become dramatised, often into end-of-the-world scenarios. The media, often not allowing accuracy to spoil a good story, have been fanning the fires. They have been ably aided by a substantial and growing number of people with an interest in capturing public attention: researchers who enjoy having their work suddenly move from obscurity to the limelight, people who organise conferences and seminars, officials who see an opportunity to enlarge their paddocks, interest groups who sense a chance to establish a power base, politicians who scramble to get to the head of the crowd, and so on. Whether the issues are real or not can be overlooked: a great new growth industry has been born.

Those who use the public concern about the environment as a power base for themselves avoid rational discussion and analysis, or a search for solutions, because their power depends on people being afraid about the future. They do not want to set sensible, achievable, standards because if these are met, their influence in the public arena will disappear.

Prospector's bough hut on the Western Australian Goldfields in the 1930s.

They have chosen the mineral industry as a target and promote the fallacy that it is one of the main threats to the environment. The facts are that the total area actually disturbed by mining in Australia at present is 150,000 hectares, or 0.02 per cent of the continent's total area. Much of this 150,000 hectares has been or will be restored after mining to the original, or even better, condition. By contrast, the area under roads in Australia is five times greater and very little of this, if any, will ever be rehabilitated.

These are the facts, but the extremists do not care for facts. They are very skilful in manipulating public opinion to suit their purposes, and are quite prepared to use distortions and untruths, and even deceit, to do so. They are also experts in fabricating what the media treat as news.

The Record

All human activities must be judged against the standards of the time. We may not agree with these standards from the vantage point of a later generation, but this is the only fair basis for assessing what was done in an earlier time. Past practices of the mineral industry, including environmental matters, must also be judged on this basis.

One early impact of mineral activities on the Australian environment came through the activities of individual prospectors in the gold rushes of the mid-1800s. This impact was through the sudden concentration of many thousands of people in the small areas of the finds, and through the disturbance of the surface while trying to find alluvial gold.

Logs at the Brilliant mine, Charters Towers, north Queensland, 1897.

While some prospectors struck it rich, most made at best a modest living. There were no legal or community requirements to rehabilitate the mined areas; such a thought, had it been put to the miners or the authorities at the time, would have been received with incomprehension,

Square set stope in a mine at Broken Hill, 1919. Although it used up timber, square set stoping was a major advance in safety.

even hostility. The alluvial mining areas were often left as unsightly dumps and holes in the ground.

This does not stop some groups today wanting to declare the early gold-rush areas a part of National Heritage, partly, it appears, to prevent contemporary mining. The eyesores have become attractive because they are 100-odd years old. By contrast, new mining projects are usually approved today on condition that everything has to be razed on completion of the project and covered over to resemble the original surface. Nobody seems able to explain this paradox.

The mineral industry is very conscious of its heritage and the need to preserve unusual and unique examples of early mining activities. But it is not rational to exclude areas the size of some European countries from future economic development. The past can be preserved adequately without penalising the present and the future.

When alluvial prospectors gave way to reef mining, the use of machinery and the underground working meant a need for wood to generate steam and to support the workings, as well as to erect buildings and dwellings and as firewood for individuals. Increasing areas around the mines were gradually denuded of trees and vegetation. In some areas, overstocking by the surrounding sheep stations compounded the problem. Choking dust storms often followed.

The first to be concerned about this and to take action were people in the industry. At Whyalla, Essington Lewis, CH, led an effort at tree planting in the 1930s, including the introduction of Athel pines, from California, which were suitable for the arid climate. In 1936 at Broken Hill, A. J. Keast, CBE, and Sir Maurice Mawby, CBE, joined the amateur botanist Albert Morris in creating, with the co-operation of the civic authorities, a regeneration zone around the town and a veritable oasis of trees, grass and recreation facilities near the Zinc Corporation mine. At about the same time at Mount Morgan, General Manager Adam Boyd and his successor Glenister Sheil led the conversion of a bleak environment into a worthwhile place to live and work. It is said that in Sheil's

Donkey team hauls timber for gold mines in the Murchison area of Western Australia. The mining industry was once a major user of timber, for lining shafts, propping roofs and fuelling steam-driven machinery. In some areas, mines removed all timber for kilometres around mining areas.

Spectacular landscape around Queenstown, Tasmania, where the countryside became devoid of all vegetation for kilometres around the mines and smelters.

time, to damage or remove a tree without permission at Mount Morgan was nearly a hanging matter. At Port Pirie in the early 1940s, A. J. Keast, with the experience at Broken Hill fresh in mind, made sure that housing areas being built there had ample provision for trees, resulting in a leafy suburb any city would value.

Some early effects on the environment could not be corrected. At Mount Lyell, the pyritic smelting introduced in 1896 to cope with the gold-copper sulphide ore caused, in that wet climate, the denuding of the nearby hills of all vegetation. This would not be allowed today, but was not unacceptable then. The multi-coloured bare hills, not unlike the famous Painted Desert in Arizona, in fact became a strong tourist attraction in this area, which is otherwise mostly covered by virtually impenetrable forest. By the early 1990s, some local people were concerned that the re-growth now starting on the hills would destroy this attraction.

Closer to the present, the preservation of native trees and shrubs and the attention given to the environment by such developments as Kambalda nickel, Pilbara iron ore, Groote Eylandt manganese, Mary Kathleen uranium, and so on, is well known to those in the industry and to serious environmentalists. The common thread in all these projects is that they were constructed well before environmental issues became a public or government concern.

Forest rehabilitation on former bauxite mined areas of the Darling Ranges near Perth, Western Australia.

Australia's aluminium industry, which began to develop before environmental issues became popular, has a proud record. As the best known example, Alcoa of Australia Ltd was in June 1990 entered in the United Nations Environmental Protection Agency's Global 500 Roll of Honour for its rehabilitation of mined areas and environmental care in alumina refinery operations in Western Australia. Alcoa was the first Australian company and the first mining company in the world to be recognised in this way. This contrasts with the extremist-led campaigns against bauxite mining in the Darling Range in the early 1980s, which claimed that regeneration would not work. The opposition stopped when Alcoa began showing Perth people and visitors regeneration areas. By the 1990s, some 25,000 people a year were seeing it themselves, and the earlier scare campaigns were a thing of the past. There was a lesson in this for the mineral industry generally: it needs to encourage the public to see what it is doing.

Rt Hon Sir Ninian Stephen, AK, GCMG, GCVO, KBE, Ambassador for the Environment (left) and Sir Arvi Parbo, Chairman of Alcoa of Australia Ltd, on the occasion of Environment Day 1990 when Alcoa was listed on the 'Global 500 Roll of Honour' by the United Nations Environment Programme for its twenty years of achievement in rehabilitation of its bauxite mined areas in Western Australia.

Visitors to the Gold Coast nowadays enjoy the sun, sand and surf unaware that large portions of the coastal sand dunes have been mined for minerals. Sand mining began along the east coast of Australia in 1935. There is now little evidence of such activity, particularly of more recent mining when rehabilitation techniques have been refined. Regenerated strip mining areas in the coal-rich Bowen Basin are hard to detect, except that the vegetation is often better and more attractive than in the unmined areas.

The Australian Mining Industry Council appointed its first full-time environmental officer in 1969 and established its Environmental Quality Committee in 1970. It organised an annual Environmental Workshop at which practitioners from industry, government, consulting, and academia met to review progress and discuss national and international issues. Typical of the topics were those listed for the 15th Workshop in Wollongong in October 1990: Integration of Mine and Environmental Planning, Mining, Conservation and Land Use, Water Monitoring and Control, Environmental Auditing, The Industry and the Community, Rehabilitation Criteria, and Rehabilitation Techniques. The Australasian Institute of Mining and Metallurgy has adopted an Environmental Policy.

All mineral operations of any size employ professional environmental staff. At large operations this can amount to many tens of specialists, such as botanists, zoologists, entomologists, meteorologists and radiologists.

Checking a flow meter on a retention pond at Ranger uranium mine in Northern Territory. All rainfall from disturbed areas on the mine site is collected in four retention ponds and handled according to strict procedures. Ranger has 35 environmental officers, of whom 24 have professional qualifications. The care of the natural waterways around the mine is a vital part of their task.

Future Outlook

While care of the environment is now a well established and integral part of all mineral operations, the techniques and procedures, as for all technology, are constantly improving. Research into environmental matters is done by the environmental personnel at operations and in co-operation with universities, the CSIRO, appropriate government authorities, and others.

An area where much improvement is needed is in fundamental understanding about society's relationship to the environment. Our very existence has an inevitable effect on the natural environment. The issue is not how to preserve all nature in a pristine manner, apart from limited areas set aside as virtual museum pieces. The issue is how to secure our economic well-being in an acceptable and sustainable way.

We also need to understand that there is no such thing as a risk-free existence. Everything we do has a risk. The realistic aim is to reduce

In the nursery at the Comalco bauxite mine, Weipa, North Queensland.

In 1989, mineral explorers and farmers agreed on a Code of Conduct for mineral explorers. Reg Fraser (right) Chairman of NSW Chamber of Mines and Michael Tooth, President of the NSW Farmers' Association.

Regeneration employees harvesting cashew nuts in the experimental plantation at Weipa. The nuts are collected to evaluate individual tree yields.

the risks, consistent with acceptable cost in economic and loss of freedom terms. The environmental area is no exception.

It appears that, just as many think that to get something from 'the Government' is free, environmental benefits are also assumed to be free or at someone else's cost. Measures to protect the environment have in many cases a very considerable cost. This cost is ultimately carried not by someone else but by all of us. Rational consideration of environmental issues and remedies must ensure that we are willing to pay the cost to secure the benefits.

Some improvements can be obtained at modest cost. We are a wasteful society, as anyone coming home from shopping can readily confirm by the pile of discarded wrappings and packaging. There is a very strong case for less waste, less unnecessary consumption, and more recycling. The mineral industry in its own activities strives continuously towards these goals, including encouraging recycling of the materials it produces.

The message I want to leave with you is this: the growth in the world population and their desire for higher living standards mean that there must be more economic development. It is not possible to stop it. But it is possible and must be our objective to pursue this development in an environmentally sustainable and acceptable way. There is no conflict between economic development and satisfactory care of the environment. On the contrary, a strong and prosperous economy is necessary for us to be able to meet the cost of environmental protection.

We need more rational consideration and less publicity hype in discussing environmental issues. These matters are too important to be dealt with emotionally, or to be decided on purely short-term political grounds.

This paper was presented in Brisbane on 24 August 1990.

Inside the wind profile building at Mount Isa Mines Limited. The building and equipment were established in 1989 as part of the Company's air quality control program.

The Mining Heritage of Central Queensland

By 1990 the Mount Morgan mine, after 108 years, really was almost at the end of its life, having been prematurely dismissed as 'finished' in 1927 and again in 1975. It had been central Queensland's first major industrial operation. For over 80 years, until Queensland Alumina at Gladstone and the great central Queensland coal mines were developed, it shared with the meat works and the railways the distinction of being among the biggest employers.

The Mount Morgan story tells of its national and international significance since its discovery in 1882. Its operations changed Rockhampton from a struggling regional port to a substantial centre of population. In 1908, it co-founded the Electrolytic Refining & Smelting Company of Australia Ltd (which became Southern Copper Ltd) at Port Kembla, for a substantial part of its life the largest electrolytic refinery in the British Empire. It developed the Dawson coal field, providing the mine with coal until the conversion of the smelter to oil firing in 1968. It helped to set up a salt works in the Fitzroy estuary, salt from which is now being used by ICI at their Yarwun plant to make chlorine and caustic soda. Its dividends to one of its major shareholders, Walter Hall, funded the establishment of the Walter and Eliza Hall Trust for charitable work and the foundation of the Walter and Eliza Hall Medical Research Institute in Melbourne. Mount Morgan Gold Mining Company Limited's largest shareholder, William Knox D'Arcy, used his profits to discover oil in Persia, an event which led to the establishment of the Anglo-Persian Oil Company Limited, which became British Petroleum Limited.

William Knox D'Arcy, the Rockhampton solicitor who made a fortune from Mount Morgan gold and took his profits to England. He almost lost them in the search for oil in Persia that led in time to the formation of British Petroleum.

Mount Morgan became a world leader in mining and metallurgical technology. For a time it had the largest gold chlorination plant in the world. It used pulverised coal to fire its power station boilers 30 years before the power industry did so. In 1972 it installed the first Outokumpu Flash Furnace in Australia. In 1982 it built what at the time was the largest carbon-in-pulp tailings retreatment plant in the world.

An early manganese mine at Auckland Point, near Gladstone.

Despite all these achievements, the Mount Morgan Gold Mining Company Limited and Mount Morgan Limited, which had the potential to be numbered amongst Australia's greatest companies, were denied such eminence by a series of crises in the 1920s, by the discovery of copper at Mount Isa, and by the company's apparent reluctance to extend beyond central Queensland. It abandoned the Blackwater coal field in the early part of the century, and did not proceed with interesting prospects which its subsidiary the Mount Morgan Developments company had across Australia (including the Tennant Creek area) from 1934 to 1938. It tried many times to establish a pyrite, sulphuric acid and fertiliser industry but was regularly defeated by fluctuations in the world price of sulphur,

Early stage of the open-cut mine at Mount Morg
Timbering of the former underground mine has been expo

and in the 1960s it abandoned a project to make calcium carbide from local coal and limestone.

How it Started

Central Queensland was settled for its pastoral potential, but mining gave it the first great leap in population.

It was from Gladstone, then the most northerly settlement in the colony of New South Wales, that the first attempt to explore the mineral wealth of central Queensland was made. Gladstone was settled in 1848, more than a decade before Queensland became a separate colony. In 1857 Maurice O'Connell, the Resident Land Commissioner, writing to the Chief Lands Commissioner in Sydney, said of a gold discovery at Canoona Station, near Rockhampton: 'I am fully persuaded ... that there is a rich and extensive goldfield in the neighbourhood'.

Early days at Mount Morgan gold mine.

By July 1858, the news that a shepherd had found a 12-pennyweight nugget spread quickly and from September to November, an estimated 15,000 men rushed to Canoona. The field had been good enough to support about 100 men making 30 shillings a day before the rush started, but not good enough to support the thousands who came.

As we all know, miners do not discourage easily and only three years after Canoona, copper was found at Peak Downs. Queensland entered the base metal industry with the formation in 1862 of the Peak Downs Copper Mining Company, which smelted 100,000 tons of 17 per cent ore between 1863 and 1867, at the same time shipping quantities of higher grade ore to Swansea, in Wales.

In the next 20 years, the areas around Rockhampton saw numerous small gold shows, some reasonably good, others short-lived. But while many wandered the gullies below Ironstone Mountain (as Mount Morgan was then known) trying to trace the origin of the alluvial gold found there, no one realised its wealth.

Although several people had thought the top of Ironstone Mountain contained gold, they had been dissuaded by 'experts' who told them, 'it's only mundic'. One man, William McKinley, knew it was gold, but took only what he needed to supplement his wage as a stockman at Calliungal Station, on which part of Ironstone Mountain stood. McKinley's daughter, who knew the mountain's secret, was married to Sandy Gordon.

Gordon's taste for tippling led to his being sacked by his employers, the Morgan brothers. His wife bargained with Ned Morgan to re-employ him, with the promise that he would show Morgan a rich silver deposit near Ironstone Mountain. Morgan agreed, but on a prospecting trip to the area in July 1882 Gordon was unable to produce any silver. When storm flooded the creeks and prevented the party returning to Rockhampton, Morgan wandered to the top of the mountain and took samples of what he was sure was gold.

Credit for the discovery of Mount Morgan's gold has long been disputed. Mount Morgan's historian and mining engineer B. G. Patterson said, [the Morgans] 'fully deserve their good fortune, but they do not deserve the credit for being the ones who really discovered Mount Morgan ... its discoverer was William McKinley'.

Underground miners at Mount Morgan using air-operated rock drills to prepare holes for blasting.

The Morgans, although experienced mining men, seemed unsure of the potential of their find, so they decided to reduce the risk by inviting Tom Hall (then Manager of the Queensland National Bank at Rockhampton) to buy a half share for £1,200. Hall agreed to a later revision of the price to £2,000 and brought in his brother Walter Russell Hall, Rockhampton solicitor William Knox D'Arcy, and grazier William Pattison, as equal partners. The £500 each of the partners put into the venture seems to have been the only capital these four ever risked, with all future investment funded from profits.

The early days of the 'Syndicate' led to formation of the 'old company', the Mount Morgan Gold Mining Company Limited, in 1886. There are no records of the profits made by the 'Syndicate' from 1882 to 1886, but they were considerable.

The 'subscribers' to the Mount Morgan Gold Mining Company Limited were the vendors of the mine and included the original syndicate members D'Arcy, Tom Hall, Walter Hall and William Pattison, who between them held 76.6 per cent of the 1,000,000 one pound shares. Despite early problems in getting reasonable recoveries, the development of the Hall-Richard chlorination process by Wesley Hall and his metallurgical engineer, Captain G. A. Richard, ensured that dividends from November 1886 to June 1892 were 347 per cent. In 1898, the Company paid £1,100,000 in dividends in 11 months.

Walter Hall (left) and R. G. Casey, Directors of Mount Morgan Ltd in 1890.

The People who ran Mount Morgan

Like most mining operations, the history of Mount Morgan is probably best told through the people who were associated with the place, beginning with the Halls and William Knox D'Arcy, who dominated the first 30 years of the old company's life.

Tom and Walter Hall came to Australia in the early 1850s, lured by the chance to make a fortune in gold mining. Both were at Ballarat, and Walter was at the Eureka Stockade. He was standing near Peter Lalor when he was shot, and his necktie was used to bind Lalor's arm wound. Walter's first fortune came from his interest in the Cobb and Co. coach line and his second from Tom's invitation to join him in the Mount Morgan Syndicate. His money stayed in Australia and on his death formed the basis of the Walter and Eliza Hall Trust, devoted to charitable works and providing the seed money for the start of the Walter and Eliza Hall Medical Research Institute in Melbourne.

Mount Morgan Ltd Directors. Kelso King (left) and Captain Richard in 1890.

Wesley Hall was General Manager at Mount Morgan from 1886 to 1891 and only he of all the Halls became involved with the Mount Morgan community. He became the town's mayor, donated the money to build Saint Mary's Anglican Church, was foundation treasurer of the first Masonic Lodge, and established the Wesley Hall Cup for soccer, which is still played for in Rockhampton.

Of the owners of the mine, William Knox D'Arcy is the least esteemed in the town of Mount Morgan, but the most famous internationally. He took his profits to England in 1886 and became one of the social lights of the late Victorian and early Edwardian era. He is best remembered for his foray into the exploration for oil in Persia, a project which almost cost him what remained of his Mount Morgan fortune but which, being successful, gave him a second windfall. It also provided Britain with a strategic advantage in the First World War, when the navy was able to fire the fleet's boilers with oil instead of coal. He founded the Anglo-Persian Oil Company, which in time became British Petroleum.

Captain G. A. Richard, who was President of The Institute in 1899 and again in 1910, came to the Mount in 1884 as an engine driver, although he was a graduate of the Ballarat School of Mines. His ability was recognised by Wesley and Walter Hall, and he became General Manager in 1904. He led the company's change from gold to copper production, which involved major mining and treatment plant changes.

Glenister Sheil, General Manager at Mount Morgan, 1960.

The years of the First World War, with its labour shortages, were followed in the 1920s by years of low copper prices and growing industrial unrest. In 1921 Mount Morgan closed for the year and in 1925 a major industrial upheaval ended in a fire which destroyed the underground mine.

In 1927 the company was placed in voluntary liquidation. In 1929 a new company, Mount Morgan Limited (always referred to as the 'new company') was floated, led by the last General Manager of the old company, Adam Boyd. The company kept afloat during the next three years (in the middle of the depression) while precipitating copper from water pumped through the old underground workings. By mid 1932, when the battle was almost lost, the Commonwealth and State Loans Council, in a programme to help the gold mining industry, lent the company £15,000 for 'test work'.

Brian Lennon had already done test work with the aid of American Cyanamid and Mount Isa Mines and after a few weeks of pilot plant work, the Government's representative Herbert (later Sir Herbert) Gepp authorised the starting up of the 'Big Mill' in October 1932. The success of the operations resulted in an infusion of capital and the repayment of the government loan by February 1933. Local legend has always had it that this early repayment, which surprised the Queensland Mines Department, was made not only to improve the company's image with shareholders but also because Boyd was anxious to stop Gepp 'looking over his shoulder' as government representative.

Capital expenditure in the late 1930s modernised the mine and plant and put the company in a good position to survive the war years that followed. While not as glamorous, the performance of the new company compared well with that of the old. From 1932 until 1972, when Mount Morgan Limited's results no longer appeared separately in the statements of its parent company, Peko Wallsend Limited, the Company averaged 15.6 per cent per year return, with a loss in only one year.

Perhaps Adam Boyd's greatest achievement was the setting up in 1933 of a welfare scheme for the employees. The Works Welfare Scheme, administered by elected employees and a lesser number of staff, received 7½ per cent of the ordinary dividends with 60 per cent for distribution as a cash bonus and the remainder to provide sporting and recreation facilities as well as a free insurance scheme.

In retirement Adam Boyd remained on the Board, as Chairman from 1938 to 1941 and as a Director until his death in 1948 at the age of 83 years. He had proposed that Mount Morgan Limited, as the largest

Mount Morgan treatment plant.

Leichhardt Colliery, near Blackwater, central Queensland.

copper producer in Queensland, should build its own refinery, but the emergence of Mount Isa as a major copper producer dashed those hopes.

Adam Boyd was awarded the Institute Medal in 1941, 'In recognition of his services in the development of underground mining methods at Broken Hill, coal mining in New South Wales, gold and copper mining at Mount Morgan, and generally for outstanding contributions to mining practices over a period of fifty years'.

Boyd encouraged concern for the environment. The timber getters of the 1880s and 1890s, supplying about 4,000 tons of firewood a month and up to 4,000 cubic metres of sawn timber a year, denuded the hills, the fumes of the smelter killed the rest of the trees, and the goats and domestic firewood collectors made sure the timber did not regrow. Boyd started a programme of timber regrowth and this was maintained by his successors, in particular by Glenister Sheil, General Manager from 1950 to 1964.

Sheil also campaigned to modernise and beautify the town which until his time had no water supply, few bitumen roads and little pride in itself. With the company providing technical assistance, the town soon had a water supply, bitumen roads, night tennis courts and a new golf course. Bank loans were guaranteed for employees to build new homes while interest free loans were provided for townsfolk to paint their premises and to install septic systems.

Adam Boyd is reported to have said in response to a request from a staff member for a rise in salary 'the climate is worth two quid a week'. Now, in addition to its excellent climate and its ideal location so close to the coast, the town was transformed from a dingy place to somewhere worthwhile to live. Glenister Sheil also worked hard to step up an already good technical training programme.

Along with Ian Morley ISO, J. Malcolm Newman, CBE, (long-time Chairman of the Company and recipient of the Institute Medal in 1961),

Julius Kruttschnitt and Glenister Sheil were instrumental in encouraging the establishment of the Mining Department at the University of Queensland.

Julius Kruttschnitt, the American who came to Mount Isa in 1930 and made his mark in Queensland mining circles, was later a Director of Mount Morgan Ltd and President of The Institute in 1952. Sheil was also a Councillor of The Institute for many years and President in 1962.

Many other distinguished members of The Institute have been associated with Mount Morgan. One was Henry St. J. Somerset who became President in 1931. His son Sir Henry Somerset, CBE, who was born in Mount Morgan, was President in 1958 and again in 1966, and was elected to Honorary Membership in 1980. J. Malcolm Newman, CBE, received the Institute Medal for 1960.

In June 1958 a merger was negotiated between Mount Morgan Ltd and Peko-Wallsend Investments Limited. Peko's policies enabled the then General Manager, F. L. (Loy) Hennessy, to preside over what was probably the company's largest capital investment programme since its seven-year programme of the late 1930s.

The final days of mining at Mount Morgan were guided by Ossie Wilson, a third generation Mount Morgan employee. Mining finally stopped in 1981, after the mine had produced 225 tonnes of gold (60 per cent of which was won by the old company) 360,000 tons of copper (30 per cent by the old Company) and 50 tonnes of silver.

For the 'last stand' operation, a 10,000 tonnes per day carbon-in-pulp tailings retreatment plant was built to retreat the 32,000,000 tonnes of tailings (averaging 1.08 grams of gold per tonne) which had been produced from the mining operations of both companies. At the time this was the largest plant of its type in the world.

Aluminium ingots being shipped from Boyne Island smelter, near Gladstone, central Queensland.

The End of the Story

The Mount Morgan mine provided employment for generations of families and made significant contributions to the technical and economic development of Australia and beyond. It will leave little behind in Mount Morgan to remind anyone of its existence. Its contribution to Queensland and Australia, however, has been tremendous.

One last contribution was the decision by Peko-Wallsend Limited, supported by the then owners, Elders Resources, to donate to the University College of Central Queensland, Rockhampton, the archival material of the mine, spanning its entire life. These archives, preserved in the Capricornia Collection within the library of the college, had been carefully guarded and nurtured.

The preservation of the history of Mount Morgan in this manner was a fitting conclusion to one of the most important chapters in Australia's mineral heritage.

I am indebted to Ray Boyle, former Chief Engineer of Mount Morgan Limited, and Gilbert Ralph, MBE, for their invaluable assistance in the preparation of this paper, presented at Emerald on 10 August 1990.

Coal from the Bowen Basin

The development of the Bowen Basin, resulting in a major export industry for Queensland and for Australia, is an example of the value of foreign investment. And there is a moral in that.

In Australia, community attitudes towards foreign investment wax and wane. In the 1960s foreign investment was welcome. In the early 1970s, when Australians thought they would ride forever on the wave of prosperity generated by mineral and energy developments, foreign investment became closely scrutinised, if not unwelcome. More recently, foreign investment again came to be seen as desirable and necessary, although with concerns expressed at times about its extent.

Coal mine at Howard, 1920s.

Since the 1960s, the Bowen Basin has been transformed into one of the top coal-producing regions in the world. Much of the development took place at the time of misgivings about foreign investment. It was argued that most of the profits from mining would be repatriated back to the home country of the company involved. This kind of thinking dogged the Utah Development Company in particular, and persisted in spite of all attempts by the company to point out that it was in error. In retrospect it was clear that Queensland and the Australian community benefited most from these developments. The majority ownership of the coal operations later reverted to Australians when BHP acquired Utah's Australian coal interests.

Early Days in the Bowen Basin

Large scale development of the coal industry in the Bowen Basin began more than a century after coal was first discovered there. The explorer Ludwig Leichhardt first noted coal deposits in 1844-45. The geologist Richard Daintree, who was also an expert photographer, in 1866 noted that:

> The Bowen River Coalfield is of great extent and contains numerous seams . . . its similarity to the Newcastle coalfields of New South Wales being to a certain extent a guarantee of its mineral wealth.

Panel on a sculpture of the city's history at Ipswich, in southern Queensland, a major centre of Queensland coal mining until the Bowen Basin displaced it.

In 1872, Daintree inspected and mapped the coal field quite thoroughly. This led to some interest in developing the field, but nothing came of it. In 1913 the Bowen River Coal Prospecting Syndicate had a tunnel driven on the Garrick Seam and sent coal samples to Bowen, where they were tested for steaming by the Railway Department and at the Merinda meatworks. The test by the Railways, on the Bowen-Proserpine Tramway, was 'a most stringent and severe one, [after which] all were most surprised at the excellent and satisfactory result'. Despite the test showing the sample to be very good steaming coal, further progress toward developing the field was slow.

Underground coal miners at Howard, in south-east Queensland. Mines here were economically viable because they were close to the main centres of population, and supplied coal for railways, industry, and electricity generation.

Elsewhere in the region, coal was often discovered by chance: in a well, river bank or shaft. Small-scale underground mining took place at various places, but lack of transport meant that only those mines producing for local use survived.

Blair Athol is a good example. The coal mine there was developed in 1890 to supply the Peak Downs copper mine, discovered about 20 kilometres away in 1861. By the time Blair Athol came into production, the copper mine had failed, so the coal was sold to the Central Railway. The cost of hauling coal to the railhead at Clermont was 'fearsomely high and accounted for about half of the contract price of 21s per ton'. As building a tramway to the railhead was not economically feasible, the mine closed in 1896.

Marking Time

By the start of the First World War, mining in Queensland was in decline. Most of the northern goldfields were worked out, leaving only a cluster of small coal fields at West Moreton in the south and the copper mines of the far north-west.

The Ryan Government, elected in 1915 and seeking to boost the mining industry, purchased the Bowen and Warra coalfields, but no development could take place there without transport. Some progress was made in 1916, when a rail link between Bowen and the coal field enabled the development of two coal mines. A private group established the Bowen Consolidated Mine, which began production in 1924 and, after extensive drilling, the site for the Government-owned State Coal Mine was eventually selected at Collinsville in 1918.

In the 1920s, Collinsville was the only state-owned mine not operating at a loss. The State Government's policy of ambitious state capitalism had included a large investment in the Chillagoe copper mines, and a proposal to establish an iron and steelworks at Bowen, using coal from

Collinsville, but the Government was unable to raise the funds it needed for this programme.

On the whole, lack of capital investment and government indifference saw mining languish in the inter-war period. Until 1940 coal production in the state remained static at around 1,000,000 tons a year. The concentration of small coal fields in the West Moreton region continued to be the most productive, benefiting from their proximity to the railways and factories of Ipswich and Brisbane.

Exporting Coal

After the Second World War, demand for Queensland coal grew in the southern states of Australia, where manufacturing was concentrated. This led to renewed interest in the industry and large-scale open-cut operations began on the Blair Athol field in 1946, where Thiess Brothers began contract mining for the Blair Athol Coal and Timber Company. This was the first time open-cut mining for black coal was tried on a large scale in Australia.

Miner underground at Collinsville Coal Company where coal seams average six meteres in thickness.

Thiess Brothers soon began operating its own mine on the south-eastern fringe of the Bowen Basin, opening the Callide coalfield in 1948. The company later expanded further, mining reserves near Emerald, Springsure and Biloela, then the largest undeveloped coal-bearing region in Australia. But the underground mines at West Moreton remained the most productive in Queensland at this stage, with growth of the industry in the Bowen Basin still constrained by lack of transport and port facilities.

By the 1950s, with the railways changing to diesel locomotives, the main use for coal was for power generation. In Collinsville, this shift was reflected in changing ownership of the mines. Mount Isa Mines Ltd,

Early open-cut coal mine at Blair Athol. Open-cut mining was introduced at Blair Athol in 1932, but abandoned in 1936. It resumed in 1946.

a major customer of Bowen Consolidated Coal Mines Ltd, had by 1952 acquired control of the company. The State Mine leases were acquired by Dacon Collieries Ltd in 1961.

Rapid Development in the 1960s and 1970s

The real boom in coal mining in Queensland came some time later, when companies such as Peabody and Utah entered the industry. In 1963, Thiess formed a consortium with Peabody Coal (then the largest coal mining company in the USA) and the Japanese trading house Mitsui, & Co. Ltd negotiating a seven year contract to supply 3,000,000 tonnes of coal per year to the Japanese market. They developed the Moura-Kianga mines to supply this.

Coal mining in Australia changed character. Peabody brought both foreign investment and expertise into the consortium. Open-cut operations on a massive scale were possible, and capital was available for the latest mining technology. Operations at Moura-Kianga were

CRA open-cut thermal coal mine at Blair Athol, Queensland.

managed by Peabody's mining engineers. At Moura, where open-cut operations began in 1961, three of the largest draglines in the world operated day and night, seven days a week.

These developments had a profound impact on the central Queensland economy, in part reshaping the region. For instance, as Ross Fitzgerald has pointed out in his *History of Queensland: From 1915 to the Early 1980s*, the choice of Gladstone rather than Rockhampton as the coastal terminus of the rail link from Moura 'marked the shifting economic fortunes within the region'.

Shovelling from the coal face, Saraji, central Queensland, in the 1970s.

At first, the State Government planned that developers would build and operate their own railways. When discussions regarding the development of Moura-Kianga first took place, the state rail system was in crisis, and the Government did not want to add to its burden. But the Government later saw the railways as a means of extra revenue-raising. Several years later, in a revised agreement with Thiess Peabody Mitsui, it agreed to build and operate the railways, but with money provided by the companies. Freight charges and a significant profit component could offset losses on other lines. While good for the Government, this was to be a major burden on Queensland coal in competitive world markets. This was in contrast to Western Australia, where mining companies built their own railways and achieved some of the lowest rail transport costs in the world.

Utah Development Company

Central Queensland coal mining took another leap forward when the Utah Development Company entered the industry. Utah's success and profitability eventually so dominated public perceptions of the company that its achievements were almost entirely overlooked.

Stockpile of coal at Saraji, central Queensland, 1975, awaiting export.

Utah's exploration began in 1961 in the Moura-Kianga region, south of the Thiess Brothers operations, and was headed by Don King. Although the team's success in finding coal was not a lucky strike by an amateur, a major discovery is always cause for celebration. Adrian McGregor's description of King's discovery of the main lower seam of coal at Blackwater in 1962 (quoted in Brian Galligan, *Utah and Queensland Coal*) conveys the elation felt by the team:

> Don King is in the usual mould of hobnailed geologists, level-headed, not excitable. But, as surely as if oil had gushed from his hard hat, this was the moment of discovery. And he knew it. He raced back into Blackwater to telephone his employers, Utah Construction and Mining Company, of his find. He sideswiped a farm gate and dented the jeep. He careered off another fence post. King had found for Utah the gently sloping sides of the Bowen Basin coal seam.

Blackwater was in those days a tiny settlement of a dozen houses, near where Leichhardt had originally noted coal deposits by the Mackenzie River, and there were not too many phones available for King to use!

In 1965 Utah signed a 10-year agreement to supply Japan with 13,500,000 tonnes of coking coal, 'the largest purchase of coal ever made

To help counter unfavourable comments about its foreign ownership, Utah publicised the benefits it brought to Australia, as in this 1977 advertisement.

by the Japanese steel industry' and the Blackwater mine was developed to meet this. The town grew quickly as production at the mine expanded, particularly following Utah's first shipment of coal to Japan in 1967.

Benefits accruing to the State from the Blackwater mine went beyond rail freight charges. When the mine was being developed, steaming coal had little commercial value. However, the Government required Utah to save and stockpile any steaming coal it removed in mining coking coal, to use for power generation. A large power station was later built at Gladstone to capitalise on what was to the Government a cheap resource.

Utah's exploration did not stop with the discovery at Blackwater. The team began searching further afield. They made geological maps of the northern area, and soon found other deposits of coking coal. In October 1962, King reported that 'the Goonyella locality gives promise of an important coal discovery'.

In 1963, Utah was granted Coal Prospecting Areas extending through German Creek to Norwich Park and Goonyella, in what was to become known as the 'Golden Mile' of the Bowen Basin. In September, King recommended a programme of drilling, targeting Booroondarra, Leichhardt Downs and Winchester Downs, the sites of German Creek, Norwich Park and Peak Downs mines, and included Saraji and Goonyella. Several of these areas were destined to become highly productive mines.

Central Queensland Coal Associates

Developing these Bowen Basin mines was a huge job, the largest project ever undertaken in Queensland. Blackwater had some basic infrastructure in place, but the more northern parts of the Bowen Basin did not.

For Utah, too, it was a massive project. A small organisation, Utah had to borrow much more than its net worth to fund the Bowen Basin mines, with development capital raised on the basis of sales contracts. In 1965 Utah formed Central Queensland Coal Associates (CQCA) with Mitsubishi Development Pty Ltd, to ensure entry into the growing Japanese market. CQCA's predominance in the industry was assured in 1968, when it took up four leases covering Goonyella, Peak Downs, Saraji and Norwich Park, and the as-yet-undeveloped Daunia deposit. That same year, Queensland achieved record coal production of 6,500,000 tonnes. With more than 60 per cent of production coming from foreign-owned operations, the question of foreign dominance of the industry began to arise.

The first mine to be developed, Goonyella, was part of a $1 billion agreement to supply 14 Japanese firms with 85,000,000 tonnes of coking coal (from Goonyella and Peak Downs) over 13 years. A new township, Moranbah, was established mid-way between Goonyella and Peak Downs and the two mines linked by rail to the new port of Hay Point. Production began at Goonyella in 1971 and by 1990 capacity was over 5,500,000 tonnes per annum. Mining began at Peak Downs the following year, the mine having a maximum production capacity of 5,400,000 tonnes a year.

Giant draglines remove overburden to expose the coal.

In 1972 coal production in the State rose a staggering 54 per cent to 17,400,000 tonnes, and Queensland coal exports surpassed those of New South Wales for the first time. In the decade from the late 1960s, the price of coking coal increased five-fold. This coincided with another of CQCA's mines, Saraji, coming into production in 1974. Utah's profits naturally rose, and it became the largest profit earner in Australia, further fuelling disquiet about the level of foreign ownership.

The Mid-1970s and Beyond

Although Utah was the best-known company, others were establishing themselves in the export coal industry. In the mid-1970s, the Collinsville Coal Company was formed by a merger of Dacon and Bowen Consolidated and in 1977 Mount Isa Mines Ltd (MIM) became sole owner of the merged company. It set about developing Collinsville, together with Oaky Creek and Newlands, and the port at Abbot Point, in a major move into the coal export business. The Newlands, Collinsville, Abbot Point (NCA) project cost some $700,000,000 and was, until the BHP acquisition of Utah, the largest coal-financing undertaken in Australia.

In 1976 coal became Queensland's major export, surpassing wool for the first time. Total exports of coal, coke and briquettes in the 1970s were valued at more than $900,000,000 a year. By the mid-1970s the coking coal boom began to subside, but demand for steaming coal increased. This shift in demand led to the development of a number of steaming coal projects in the late 1970s and early 1980s. CQCA however was prevented from exporting steaming coal as part of its 1968 agreement to provide steaming coal for power generation.

Coal train on the Goonyella Line, central Queensland. The Queensland Government saw the coal-export railways as a means of revenue-raising. It agreed to build and operate the railways, but with money provided by the companies. Freight charges and a significant profit component helped offset losses on other lines.

Gladstone power station.

Output grew rapidly in the 1980s, as mines such as Gregory, Norwich Park and the open-cut mines at Blair Athol and Oaky Creek came into production. By 1983-4, Queensland was producing half of Australia's coal exports. At the same time, a major shift in ownership of coal mines in the Bowen Basin took place with BHP's acquisition of Utah, which was completed in 1984. The down-turn in the industry in the 1980s was severe, and resulted in a number of closures of less-efficient mines. It also led to major re-structuring in the industry.

In 1990, coal production in Queensland rose 13 per cent to 74,100,000 tonnes. Exports reached just over 59,000,000 tonnes, including 40,000,000 tonnes of coking coal. Japan remained the largest export market for Queensland coal, with Japanese customers buying almost half its total shipments.

In the Bowen Basin, higher prices and productivity gains from improved work practices were expected to lead to increased tonnages of coal being mined. Two coal prospects, Gordonstone and Ensham, were likely to be developed, and expansion of existing mines was under way. Blair Athol, the State's largest open-cut steaming coal mine, would have a capacity of 8,000,000 tonnes a year on completion of its expansion programme. Capacity at Curragh was also being increased. MIM's plans for Oaky Creek were expected to double output of coking coal by the end of 1991.

The Queensland Department of Mines estimated proven black coal resources in the State in the order of 37,000,000,000 tonnes, almost three-quarters of which were in the Bowen Basin. The mineral heritage of the Bowen Basin would continue to benefit all Australians.

Newlands coal mine.

Associated Developments

One of the significant flow-on effects of the growth in the coal mining industry in central Queensland was the emergence of Gladstone as a major industrial centre.

Gladstone power station was built to take advantage of the steaming coal acquired at cost under Utah's Blackwater lease agreement. This in turn encouraged the establishment of Queensland Alumina's refinery at Gladstone and Comalco's aluminium smelter at Boyne Island, giving Gladstone an integrated aluminium industry.

I am indebted to Vicki Lukins and Ian Dymock of BHP Minerals for the preparation of this paper, which was presented in Collinsville on 9 August 1990.

Washing coal at Tarong, Queensland.

Our Industry's Proud Heritage

Sir George Fisher, CMG.

Ten years ago I had the privilege of presenting the Sir George Fisher Lecture here in Townsville, on the subject of 'The Nickel Industry'. Today I would like to speak about our mineral heritage. This is my general theme while visiting Branches of The Institute as President in 1990. Mineral heritage is a fitting subject for a Sir George Fisher Lecture, because Sir George is a prominent part of this heritage.

Our first thought about 'heritage' is that it refers to the past. It means the discoveries, the developments, the mines and the processing plants, the benefits, the consequences, and the human experiences flowing from these.

But there is also a much wider meaning to mineral heritage. Our industry's past has left us with the ability to create and operate world-scale enterprises generating wealth for the community, combined with the ability to offer satisfying and challenging careers in a wide range of disciplines to our young people. It has left us with the willingness to follow adventurous and challenging vocations in remote areas.

The industry continues to be a reservoir of the pioneer spirit and a vital contributor to improving life for all people. So, while our mineral heritage has much to do with the past, an essential part of it is how we use this rich past to build and shape our future.

The mineral industry has been one of the essential foundations of Australia's development and prosperity during the last 150 years. Sir George and his contemporaries will always be famous in the annals of mining in Australia. They were active and the leaders of the industry during a period when it was lifted to a new level of greatness. Their tremendous contribution to this country was well understood until the very prosperity resulting from it regrettably caused this understanding to be weakened, if not lost.

The vast majority of Australians, most of whom live in large cities, have never seen a mine, an ore treatment plant, a smelter, or a refinery, and do not have first-hand knowledge to test the validity of the misinformation so ably distributed by the industry's opponents. Those of us who do know the truth must therefore speak up. Most of all, we must explain how mineral production contributes so much to our living standards. Without the mineral industry, Australia would be a much poorer country indeed.

As your President in 1990 I am using every opportunity to do so. But we need more voices: the more the better. I urge all members of The Institute to speak up whenever they can, particularly to audiences not familiar with the industry. We should not think that our individual voices are insignificant and cannot influence the course of events. Public opinion is the most important force in a democratic society. Our individual voices, sounded often enough, and provided there are enough of them, will add up to a powerful chorus.

The Present

The industry and The Institute are well and active. The anti-development and anti-mining campaigns of recent times have managed to slow down mineral projects, and at times even defer them. But they have not stopped the industry's growth. New operations are under development in Australia, Papua New Guinea, and New Zealand.

Abandoned and red with rust, machinery lies silent around many former mining sites in north Queensland.

The late 1970s were a time of unbounded optimism, when the demand for mineral products all over the world seemed to be increasing rapidly for as far ahead as could be seen. Extra production capacity installed worldwide in late 1970s and early 1980s to meet the expected demand, together with a world economic downturn in the early 1980s, resulted in an oversupply of virtually every metal and mineral. Prices fell to uneconomic levels. Some observers began to say that the advent of high technology was reducing the demand for minerals and such views created severe doubts about the future of the industry.

In fact, in the late 1980s the demand for minerals and metals, while still as subject to short-term ups and downs as it always will be, continued to increase. The rate of increase had slowed down and the consumption increase relative to the growth of economic activity was lower, but the industry had a growing market for its products.

While economic concerns about the longer-term demand for mineral products had virtually disappeared, another major challenge had arisen. Throughout the world there was now an acute consciousness of the effect of human activities on the natural environment and the various longer-term problems this might cause.

North Queensland mines were a major source of business for firms in southern parts of Australia.

The genuine desire to care for the environment and efforts to alleviate and limit the undesirable effects are sensible and welcome. We all live in this world, albeit briefly, and nobody wants to leave our children and grandchildren an environmental legacy which is unsatisfactory and could have been avoided. The mineral industry not only agrees with this but has in many instances taken the lead in environmental care and rehabilitation of areas disturbed by its activities. In some cases, this had preceded public recognition of these issues by decades.

Those who use public concern about the environment mainly to establish a power base for themselves are not interested in rational discussion and analysis of the problems and search for solutions, because their power depends on people being afraid about the future. They are not interested in setting sensible, achievable standards because if these are met, their influence in the public arena would disappear.

They have selected the mineral industry as one of their targets and are promoting the fallacy that the mineral industry is one of the main threats to the environment.

The facts are that the total area actually disturbed by mining in Australia at present is 150,000 hectares, or 0.02 per cent of the total area. Much of this 150,000 hectares has been or will be restored after mining to the original, or often better condition. By contrast, the area under roads in Australia is five times greater and very little of this, if any, will ever be rehabilitated.

These are the facts, but the extremists are not concerned with facts. They are very skilful in manipulating public opinion to suit their purposes and are quite prepared to use distortions and untruths, and even downright lies, to do so. They are also experts in fabricating what the media treat as news.

Support from public opinion is essential for the future of the industry. Without it we will not be able to continue making our vital contribution to the well-being of every Australian. One reason the extremists have been so successful is that we in the industry have not been speaking up. We have been very bad at explaining ourselves and pointing out the distortions and fallacies in what our critics say, and even worse at setting a positive climate of public opinion ourselves. We have concluded that much of what is said by our critics does not make sense, and because what does not make sense cannot be pursued far in our professional activities, we have tended to conclude that it will also disappear in the public arena.

Copper awaits shipment from Townsville.

This is a serious error, because in politics there is no such thing as an automatic rejection of what does not make sense. Great empires have been founded on false ideas. Given enough pressure from special interest groups, plus skilful media handling, what does not make sense has a great chance of becoming law affecting the industry. Much of it has.

The Challenges

Looking to the future, the industry faces three major challenges. It must regain public support. It must continue to discover ore. And it must encourage high calibre young people to train for careers in the industry.

Regaining public support is the first challenge, because it is basic to the future of the industry. When all our energy and ingenuity are required to keep this year's financial bottom line looking respectable, it may seem difficult for people in the industry to spare the time and effort for any activity which does not have an immediate commercial impact. Events of recent years should have shown beyond doubt that we have no choice in this, for if we do not have public support we will go backwards.

We must not underestimate the forces against us, masquerading as occupiers of the high moral ground but in reality interested mainly in power.

Politicians, with some distinguished exceptions, these days follow opinion polls and support popular issues. Too many of them are only concerned with gaining office and being re-elected. Many do not understand the real issues.

A survey reported in the London *Sunday Times* highlighted this, when 75 members of the British Parliament were asked five simple questions about the objectives of well-known proposals to address specific environmental problems, such as 'what is the purpose of removing chlorofluorocarbons from aerosols?' For each of the five simple questions they were given a choice of four simple answers. Only 12 of the 75 politicians answered all the questions correctly, and of the so-called 'green' politicians only three out of 30 knew the answers. Two of these 'environmental politicians' could each only get one of the five questions right. It would probably be equally instructive to conduct a similar survey in Australia.

Gold processing plant at Mount Coolon, 1930. The building was sheeted with bark as an economy measure

I cannot repeat often or strongly enough that everyone in the industry, and certainly every professional in the industry, has a responsibility to speak up and help in the mammoth task of regaining public support for the mineral industry.

Deposits of minerals such as bauxite, iron ore, and coal are usually very large and enable continuing production for decades, even hundreds of years. Other mineral occurrences, notably orebodies mined by underground methods, are more limited. But even the large deposits will have to be replaced by new discoveries in what is in historical terms a short time. It is vital for the future that all potential mineral bearing land is available for exploration. This contrasts with the trend in Australia of increasingly locking up large areas under various pretexts.

The Institute

Our Institute represents technological professionals in the mineral industry. The industry itself is represented by other organisations. The Australian Mining Industry Council (AMIC) is the Australian national industry organisation, the Chambers of Mines being its equivalent in the states. Most members of these organisations are companies, not individuals.

A part of our heritage as professionals is the responsibility to ensure that we and our successors can continue to make the great contribution to the community made by our predecessors.

In its now nearly 100 years of existence, The Institute has been a highly respected body fostering technical progress and excellence and nurturing professional and ethical standards, professional education, and the professional standing of its members. We must continue and strengthen all these activities. High professional and public standing and respect are essential to achieving our aims. We must work tirelessly towards this.

But in the changing political and community environment we must also become active participants in public policy formulation affecting our industry and members. Policies are becoming far too important to be left to politicians and public servants.

The key to being able to discharge our responsibilities is in ensuring that we are the premier professional body in the mineral industry. We must attract a high proportion of those eligible for membership as our members. While accurate numbers are not available, we believe that it is realistic to think in terms of doubling the membership to 15,000 over a number of years.

It is very important to have a high proportion of the eligible students and young graduates as our members. The present becomes the past sooner than we like; leadership of The Institute and the industry will soon be in the hands of today's students.

To achieve these aims we must work relentlessly to make The Institute relevant to its members' needs and aspirations. Greater membership will in turn make it easier to achieve this and will enable The Institute to speak with enhanced authority.

The training and education of young men and women to join the ranks of the professions has never been more important than now. We have first class people today, but the quite-literally defamatory attacks often made on the industry may mean that other occupations are coming to be seen as offering a higher standing in the community, most of all by those who live in the big cities. There is the danger that the mineral industry may not attract enough of the best young people.

The Institute is doing something about this. The AusIMM Education Endowment Fund established in 1989 is now operational and the first scholarships have been awarded.

The Fund is focussed on attracting high calibre students to study disciplines relevant to the industry and on establishing and maintaining contact between the industry and the students during their studies. It is vital that company scholarships do not stop at providing the money. Continuing liaison with the students is essential, to gain the greatest benefit. For unnamed scholarships, The Institute must fill the role.

The Fund is up and running, but the effort is limited by the funds. More are needed to build on the good beginning.

I conclude as I started: our heritage is as much about the future as about the past. Our industry and professions have the responsibility to continue being great contributors to the community in the future, as they have in the past; this is an essential part of our heritage. There are forces opposing this, which must be overcome as we overcome many other difficulties and adversities in our professional lives. Success in this is largely up to ourselves, the members of The Institute.

This paper was presented in Townsville on 8 August 1990 as the Sir George Fisher Lecture.

Sir George Fisher, CMG, (right), with Sir Arvi Parbo (left) and J. Campbell Jenkins, at the Sir George Fisher Lecture in Townsville, 1990.

Mining Brings Prosperity to North Queensland

In September 1865 the people of Townsville, eager to promote their port against the rival claims of Bowen and Cardwell, offered £1,000 reward to the first person to find payable gold in their hinterland. Two months later a station overseer named Gibson reported gold on the Star River Station, 80 kilometres west of Townsville, and a public meeting voted him £500 reward. A small rush set in.

Richard Daintree, who had worked with the Victorian Geological Survey, brought systematic prospecting to north Queensland.

Richard Daintree, who had worked with the Victorian Geological Survey, brought systematic prospecting to north Queensland. An expert photographer, he spent months photographing, amongst other things, rock outcrops. In 1866, he found and mined copper ore on The Einasleigh and carted several tonnes to Townsville by bullock team, but freight made the venture unprofitable.

Daintree advised a party of six prospectors to explore the headwaters of the Cape River and they found payable gold in July 1867. Although this discovery was overshadowed by James Nash's discovery of the Gympie field in October 1867, it marked the turning point in north Queensland mineral discovery. Other finds soon followed, such as Ravenswood by Marmaduke Curr, Thomas Aitken, and others in 1868. Daintree also explored the headwaters of the Gilbert River, where a major rush followed. This led, with other nearby discoveries, to development of the Etheridge Goldfield.

Ravenswood, because of its location and high grade reefs, soon became a prosperous centre with 2,000 people and a 10-head stamp battery. Townsville captured most of the trade even though Bowen had a better port. With the barrier of the Burdekin River crossing and squabbles amongst its businessmen, Bowen languished.

Phillip Sellheim, Goldfields Warden at Ravenswood.

Bigger discoveries were imminent. Alluvial gold on the Broughton River resulted in the discovery of the '70-mile' and Rockford fields in 1871. Then came the greatest of them all, Charters Towers, discovered by Hugh Mosman, George Clarke and James Frazer in January 1872. By the end of 1872 about 3,000 miners were on site.

Prospecting continued unabated. Phillip Sellheim, the Goldfields Warden at Ravenswood, recorded: 'If the Northern miner has one besetting sin . . . it is readiness at a moment's notice to sacrifice his all if required, to enable him to hurry off to the scene of some new discovery – good or bad, authenticated or not'.

The most romantic discovery was yet to come when William Hann led an exploration party into Cape York Peninsula in 1872, offering half a pound of tobacco to any man who discovered gold. Frederick Warner, the surveyor, claimed the tobacco when he found gold at a place they named Palmer River. Hann was not optimistic, but James Venture Mulligan and his party explored the area with success in 1873. Mulligan urged caution because of isolation and difficulties, but nobody heeded his advice, and the greatest gold rush in Australia for 20 years began.

Reacting to reports of great hardship and general distress, the Government authorised a special expedition 'to guard against the hardships which must be endured by the diggers'. The expedition left Brisbane on the S.S. *Leichhardt* in 1873 under the leadership of the Gold Commissioner, Howard St George, who had held the position at Charters Towers, the Etheridge and other fields. He was 'a man of known integrity and ability'. St George reported that the Palmer was 'the richest and most attractive goldfield yet discovered'.

During its alluvial days the Palmer was an ideal small man's field. Miners without capital or experience could get rich quickly. Output peaked in 1875 at 260,000 ounces and by 1879 over 1,000,000 ounces had been won. Although reef mining started in 1876 it was never successful and by 1892 only 1,034 people lived in the area.

The other feature of the Palmer was the large influx of Chinese. Their number increased from 7,000 in February 1876 to 18,000 by mid-1877, accounting for 73 per cent of the population. Peter Bell recorded from the Bowly Letters collected by Dr Kett Kennedy that, 'the whole district is crowded with Chinese who keep stores, butcher shops, gambling houses and also grow vegetables at every different settlement. You might also fancy yourself in China – there are so few whites to be seen'.

Elsewhere in north Queensland alluvial prospectors and miners gradually gave way to reef mining, firstly by partnership and then by companies. In the 1890s and the early part of the 1900s over two-thirds of north Queensland's gold came from the Charters Towers field. 'The Towers' was the second city of Queensland, with 25,000 people who proudly called it 'The World'. Production peaked at 319,572 ounces in 1899 and during its great days from 1885 to 1912 its annual output never

Mount Garnet mines, about 1900.

Charters Towers. The Brilliant mine, about 1897. Charters Towers soon settled down to the steady undergound mining that called for large sums of capital. Many companies were formed during the life of the field.

fell below 100,000 ounces. Then production fell away year by year to 147 ounces in 1928. The Charters Towers field, including production from the Cape River diggings, produced almost 6,500,000 ounces over its life, which effectively ended in 1914.

The Croydon field was the second largest producer, up to 1914 accounting for 1,357,000 ounces. Croydon had suffered like many other fields from inefficient working and the usual shady company promotion. Of 31 companies floated in 1877, Professor Geoffrey Bolton reports that by 1890 'twenty four were in liquidation, fourteen idle and only one of those still working was a dividend payer'.

Hugh Mosman, one of the three men who in 1872 discovered the golden wealth of Charters Towers.

As well as being third largest producer in north Queensland, Ravenswood became the centre of attention in 1891 when it used the cyaniding process for the first time in Australia. As Professor Geoffrey Blainey, AO, observed in *The Rush That Never Ended*, 'every goldfield gained from the Scottish cyaniding process which Australians first used at Ravenswood'. He goes on to record that in Charters Towers, in its peak year of 1899, more than half the gold produced came from cyaniding the discarded dumps. Ravenswood increased production after 1901, because of more efficient methods of working and management by such men as Archibald Lawrence Wilson, 'the uncrowned king of Ravenswood'. Production to 1914 totalled 962,000 ounces.

The Etheridge was regarded as a poor man's field. By 1915 it had produced only 500,000 ounces. Although its potential was extolled by promoters and mining men, Professor Geoffrey Bolton suggests it was, 'a museum of the various ways in which a London company could lose money'. Bad and extravagant management and unscrupulous promoters were frequent.

THE NEW RUSH

CAUTION !

The Police Magistrate has received instructions to issue the following notice:—

"Department of Public Mines,
"Brisbane, April 22nd, 1876.

"Reliable information has been received by this Department from the Warden of the Palmer River Gold Fields, who has just returned from a visit to the locality, that the NEW RUSH is a TOTAL FAILURE.

"All the Diggers are making their way back, and great dissatisfaction and distress prevail amongst them."

Court House, Rockhampton.
April 22nd 1876.

Printed at the "Argus" Machine Printing Office, East-street, Rockhampton.

During its alluvial days, from 1873, the Palmer was an ideal small man's field. But the field was short-lived, and the Mines Department issued notices warning that 'the NEW RUSH is a TOTAL FAILURE'.

Chinese miners on the road to The Palmer, in the fanciful view of a contemporary newspaper artist.

By the mid-1920s the fields were almost abandoned. It took another 60 years before there was a major resurgence. In the 1980s, Pajingo and Mount Leyshon on the Charters Towers field, Wirralie, Ravenswood, Red Dome near Chillagoe and Kidston on the Etheridge field became major producers, the Kidston mine having one of Australia's largest outputs.

Base Metal Discovery and Development

Few prospectors in the 1860s and 1870s were interested in base metals such as tin, lead or copper. It was the age of gold. But some men did take an interest in less immediate promises of prosperity.

In 1875, James Venture Mulligan reported tin on the Wild River. But real interest came in 1878 when a pastoralist, John Atherton, found tin on Tinaroo Creek. Atherton made another find on the Wild River near Mulligan's old camp. Although he was not interested in mining, he guided William Jack, John Newell, Thomas Brown and John Brown to the site.

Charters Towers Stock Exchange in its heyday.

Tin – John Moffat

Tin mining soon became concentrated on Herberton and Irvinebank, with production reaching 106,000 tonnes in 1881. The Great Northern became the premier mine, thanks to the management of John Newell, whose systematic exploration and development kept it productive.

In 1880 another shareholder, John Moffat, entered the Great Northern. Moffat, a Scotsman who had migrated to Brisbane in 1862 and been a storekeeper on the Stanthorpe tin field, was to become a dominant figure in far north Queensland. He was deeply religious, and his honesty and benevolence were legend. Exploration and the development of mines and their associated communities, not personal profit, was his major motivation. Irvinebank and Herberton developed as solid mining towns and, because of the self-help and determination of their citizens, had Schools of Arts with well-equipped libraries.

Charters Towers survived as a substantial city even after its golden prosperity had waned. Major buildings like the old Australian Bank of Commerce found new uses, such as accountants' offices.

Not all the mining townships had such reputations, Chillagoe was described as 'a village which might have been built at any time before Noah's Ark'.

While Moffat's main mine was the Great Southern at Irvinebank, he also developed other properties including several silver mines, and prospected for copper at Chillagoe and Mount Elliott. Although difficult economic times in 1888 forced him to temporarily close his Irvinebank mill, he persevered and later gained control of the Vulcan mine, which was to become the main-stay of his mining activities.

During his lifetime John Moffat developed a mining empire that earned more than £4 million in base metal production. His exploration and development, together with his policy of treating ores from outside mining operations, resulted in the employment of several thousand men and livelihood to their families. It is a measure of the man that, the story of the children of Irvinebank ending their bedtime prayers with 'God bless John Moffat' holds such credence today.

Copper – John Moffat and James Smith Reid

Patrick Molloy discovered copper at a place that came to be known as Mount Molloy and began to work it in 1885. John Atherton's son William had a cattle property named Chillagoe, west of Thornborough, where copper was discovered at various sites by Anthony Linedale and J. Delaney, who were prospecting for John Moffat.

In 1897 John Moffat invited two well known Melbourne mining men, Charles Chapman and James Smith Reid, to float the Chillagoe properties. They also set about the development of his freehold copper properties at Mount Garnet, which had been discovered by Albert Vollenweider in 1883.

Photographs showing gold bullion were often used in nineteenth century promotion for mining companies.

Mount Garnet Freehold Copper and Silver Mining Company was floated and the Under-Secretary for Mines later stated that 'Mount Garnet was the most foremost copper mine in the State'. But Moffat's caution in planning to use abandoned plant was overruled and the Board undertook an extravagant programme of development, including a railway spur from Lappa Junction to Mount Garnet. Some 700 men were employed, but Mount Garnet's life was short-lived. Over-capitalisation, falling copper prices, collapse of the open-cut, refractory underground ore containing zinc, drought, and an outbreak of typhoid resulted, not surprisingly, in the demise of the company in 1903.

In the period from 1890 to 1930, copper mining in north Queensland was dominated by the rise and fall of the Chillagoe field.

Mulligan and Moffat are well known names in far north Queensland: for years they both worked and lived on the mining fields and won and

Cooktown, arrival port for hopeful diggers bound for the Palmer River Goldfield. Charlotte Street, about 1874.

John Moffat became a dominant figure in tin-mining developments in Irvinebank and Herberton.

Chillagoe, about 1907. It was described as 'a village which might have been built at any time before Noah's Ark'.

lost fortunes. By contrast James Smith Reid is barely known. At first he was a journalist who spent little time in north Queensland. He was the son of a Presbyterian minister from Ireland and arrived, aged 14, in Bowen in 1863. As a young journalist and publisher he launched the *Ravenswood Miner* in 1870, the *Northern Miner* in 1872, the *Cooktown Courier* in 1874, and the *Hodgkinson Mining News* in 1877. In 1880 he started the *Silver Age* at Broken Hill, and by 1888 was a member of the Board of Directors of The Broken Hill Proprietary Company Ltd and involved in the Silverton Tramway Company and Emu Bay Railway Company.

In 1897 Reid, with Charles Chapman and John Moffat, undertook the ambitious development of the copper deposits at Chillagoe, including the mine, smelter and railway. Despite the contrary policy of the Queensland Government, Reid and his colleagues in 1898 negotiated the private development of the railway despite 'stormy opposition from the Labor crowd'. After a large expenditure on wharves, railway and mine infrastructure and development, production commenced in 1901 at the time of a major downturn in metal prices. This, combined with a dramatic miscalculation in ore reserves, resulted in the collapse of the company after only 8,129 tons of ore had been treated.

Reid reconstructed the company in 1905 and the high metal prices of 1906-07 gave a short period of prosperity to Chillagoe. To overcome the high fuel costs of coal and coke from Newcastle, Reid also developed coal deposits at Mount Mulligan. Not only did Reid have to contend with the many problems of the natural environment, but also the problems of organised labour.

Poor financial results and lack of liquidity forced a further reconstruction of the company in 1912. The *Australian Mining Standard* commented that Chillagoe was 'a sink for very large sums of money

Tin mine and treatment plant at Irvinebank.

Edward Granville Theodore. Miner, Union Leader, Queensland Premier, Federal Treasurer, Company Director. Theodore featured in the Mungana scandal, which involved the sale of the Chillagoe mines, smelters and railways to the Queensland Government.

which it is doubtful shareholders will ever have returned to them', 'a company of a thousand disappointments and many more of these are yet in store'. Between 1902 and 1914, the Chillagoe smelters treated about 600,000 tons of ore with a yield of 23,272 tons of copper, 31,758 tons of lead, 4,345,309 ounces of silver, and 28,911 ounces of gold.

After protracted negotiations, the Ryan Labor Government, which came to power in 1915, acquired the railways, smelters, the mines and mining plant and machinery. Chillagoe Limited, the largest mining enterprise in northern Australia, was left with only the small developing Mount Mulligan coal mine. The explosion that killed 75 men at Mount Mulligan on 19 September 1921 was also the death knell of the company. Reid died 16 weeks later, aged 73 years. The company was liquidated in 1923 and, sadly, all its records were destroyed. Although the Government-owned Chillagoe smelters and numerous mines operated until 1926, the whole affair became the subject of a 1930 Royal Commission which shattered the political careers of two Queensland Premiers, Theodore and McCormack.

There is no reason to doubt Reid's determination to make the Chillagoe company successful. Although the company accumulated debts of over £5,800,000, his companies prospected and developed an area over 10,000 square miles, built a private network of railways still in use today and exceeded only by those of the Pilbara iron ore mines, and generated jobs for over 5,000 people in an under-developed area of the Far North. Late in 1917, the *Cairns Post* editorialised, 'the major beneficiary of the Chillagoe company's endeavour was the wage earner'.

Up to the end of 1919 the total value of gold and minerals mined in Queensland was £130,800,000. The contribution of mining has been well summarised by Professor B. J. Dalton in *Readings in North Queensland Mining History*, in which he wrote: 'It used to be said that Australia rode on the merino sheep's back; with at least as much truth it could be said that from 1870 to 1920 north Queensland travelled in the miners' ore cart'.

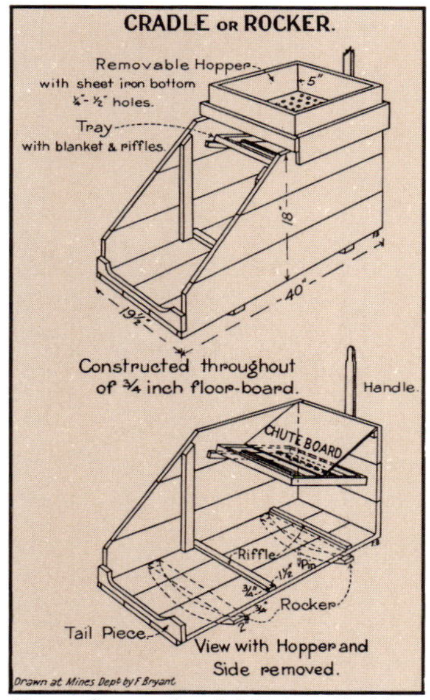

From Queensland Mining Guide 1930s.

The decline of mining in the 1920s led the Queensland Government to set up a Royal Commission, which handed down its findings in 1930. Many of them, to do with leasing, lease rents, labour conditions, tax and stamp duty concessions, lowering of railway freight rates and grants for roads and bridges, were adopted, and this did lead to more mining activity. Of particular significance was the Mount Isa Railway Act, which provided much-needed encouragement for the struggling enterprise at Mount Isa which, seven years after discovery, was not yet in production.

Queensland, like other Australian states with extensive gold mineralisation, saw a revival of gold mining during the 1930s depression. This revival followed the increase in the price of gold, because of the United Kingdom abandoning the gold standard and a 25 per cent depreciation of the Australian pound. With this increase, deposits that had been uneconomic now enjoyed renewed prosperity and once again the fields of north Queensland attracted a new wave of prospectors and a few entrepreneurs. The Second World War disrupted this regrowth and it never really revived.

The next phase of resurgence in gold mining followed the introduction of the free market price for gold in 1968 and the sudden price rises in the early 1970s, which prompted further exploration, leading to discoveries such as Kidston. Later developments included substantial gold

Stockpiling processed bauxite at Weipa for shipment to aluminium refineries in Australia and overseas.

operations at Red Dome, Mount Leyshon and Wirralie and smaller operations on the Palmer River and Charters Towers goldfields. Alluvial tin mining in far north Queensland also had a resurgence, particularly by open-cut methods in the Burdekin River area and at Mount Garnet and Ravenshoe, on the Atherton Tablelands.

Although low-grade bauxite had been identified on the west coast of Cape York Peninsula by C. F. V. Jackson, the Queensland Government Geologist, in 1902, it was not considered of any commercial significance. In 1953 Sir Maurice Mawby, CBE, who was once described as a person 'combining the optimism of a bush prospector with the caution of the natural geologist', showed some bauxite to Harry Evans, OBE, of Conzinc's petroleum exploration subsidiary and suggested he keep an eye out for it whilst looking for oil bearing structures on Cape York Peninsula. Evans saw the 'reddish cliffs' which had attracted the attention of Matthew Flinders in 1802 and many others who passed by later. Evans' systematic sampling led to the recognition of the vast reserves of bauxite at Weipa.

The subsequent development of the Weipa bauxite deposits, the Gladstone alumina refinery and the Boyne Island aluminium smelter were enormous significance to the economy of Queensland and Australia.

The discovery of lateric nickel at Greenvale in northern Queensland made 1966 a notable year for the nickel industry in Australia. The Greenvale discovery was made by Metals Exploration Ltd who, in a joint venture with Freeport Minerals Company of United States, brought the deposits into production in 1974. The ore had to be transported 225 kilometres by rail to Yabulu, north of Townsville, there to be reduced to nickel oxide and nickel cobalt sulphide, using an ammonia-ammonium carbonate leach process.

In 1989-90 Queensland produced minerals and energy valued at $4,380,000,000 and employed directly 21,135 persons.

Although the dividends from early mineral activities, particularly from gold mining, were seldom reinvested in north Queensland, mining did result in the rapid development of ports, railways and government infrastructure.

Without the mineral developments of the past and the mining and processing operations of the present, far north Queensland would be an immeasurably poorer place to live in. Most Australians know Cairns and Port Douglas as holiday and tourist destinations, but perhaps do not realise that both owe their origins to mining, as do so many other centres of population. In this, far north Queensland is a particular example of what is true of much of Australia.

While it is important to develop other types of economic activity, minerals and the rural industry will remain the backbone of Australia for decades, if not centuries, to come. In this, far north Queensland will have an important part in the future, as it has in the past.

Bauxite moves to stockpiles, Weipa, in the 1970s, ready for shipment to alumina refineries in Australia and overseas.

I am indebted to Ted Davies, former Research and Technical Manager at Mount Isa, and Gilbert Ralph, MBE, for their invaluable assistance in the preparation of this paper, presented at Cairns on 7 August 1990.

Copper and Lead from North-West Queensland

Early explorers' reports had much to say in favour of the country south of the Gulf of Carpentaria. They include John Lort Stokes, commander of the *Beagle* (who in 1841 discovered the rich grasslands of the Plains of Promise near Burketown), the Leichhardt expedition of 1845 and Burke and Wills, who in 1861 referred to the country between the Tropic of Capricorn and the Gulf of Carpentaria as 'well watered and richly grassed'. Early experience soon showed there was also a devastating downside, such as discovered with the ill-fated Burketown settlement in 1865.

The rocky, corrugated topography of the Cloncurry and Mount Isa Mineral Fields, comprising steep spinifex-covered hills and valleys, together with a harsh summer climate, erratic rainfall and drought, hostile Aborigines, and isolation, combined to make a forbidding environment for early pioneers. It was not country for the faint-hearted, but successful seekers of minerals have rarely been faint-hearted.

John Campbell Miles, the prospector who in 1923 chipped from a hill near the Leichhardt River the rock samples that led to the devleopment of the mining industry at Mount Isa.

Ernest Henry, a tough veteran of the Crimean War, arrived in Australia in 1857. As a member of George Elphinstone Dalrymple's expedition of 1859 he had explored the Burdekin Country of north Queensland and it was he who discovered the Great Australia copper deposit at Cloncurry in 1867. Later, with the help of the Kalkadoon Aborigines, Henry went on to discover the Crusader, Dobbyn, Argylla, and the Mount Oxide deposits in 1882. About the same time, William McPhail and Robert Johnson found copper at what became the site of the smelting town of Kuridala.

Following a favourable report by a Scottish geologist, J. R. Robertson, Henry, in 1884, sold the Great Australia to Scottish investors who formed the Cloncurry Mining & Smelting Company. Smelting began on 16 June 1885 and ceased in June 1887 after producing only 404 tons of copper. The Scottish company faced a major slump in the copper price from £70 per ton to £44 per ton, high wages, and heavy freight charges via the port of Normanton. The isolation and difficulties exacted their toll.

However, Cloncurry struggled and survived. In 1899, times were more favourable and the dramatic success of Mount Lyell in Tasmania encouraged a copper exploration boom. William Orr and his colleagues Herman Schlapp and William Knox, all 'Broken Hillionaires' who had also benefited from investments in Mount Lyell, purchased the Hampden property, re-discovered by Fred Gibson. The Duchess and Trekelano deposits discovered by Jack Kennedy in 1897 were sold to the Hampden group in 1906. The third major mine, Mount Elliott, was discovered by James Elliott in 1899, some 30 kilometres south of the Hampden orebody.

In the decade 1897 to 1907 the Cloncurry field revived and gathered momentum as the copper price increased from £55 to £109 per ton.

Great Australia Copper Mine, Cloncurry.

Cloncurry residents had lobbied long and hard for a railway. The Government promised to build a line from Normanton to Cloncurry and on 19 October 1886 voted the first stage. But the route was diverted to the booming gold town of Croydon, which seemed to offer more traffic. Cloncurry had to wait until a line reached it from Richmond, in December 1908. The line extended to Mount Elliott in August 1910, to Duchess in October 1912, and to Ballara and to Mount Cuthbert, in the northern part of the field in 1915.

Before the railway, transport was by horse-drawn heavy four-wheeled waggons or pack camels, thirty in a string plodding along at two miles an hour. Copper ore sent to southern smelters via Winton or Richmond required a grade of at least 30 per cent to be worthwhile. W. S. Robinson, in his memoirs *If I Remember Rightly*, recalled that in May 1906 he walked from Richmond over three days and three nights alongside an overloaded coach across the rain-soaked black soil plains, finally arriving in Cloncurry after five days.

The Rise and Fall of the Cloncurry Copper Field

Douglas MacGillivray, the Cloncurry publican and storekeeper who consolidated all the leases on the Mount Isa field.

Although some 24 companies held properties in the Cloncurry district, the field became dominated by the Melbourne-based Hampden Cloncurry Copper Mines Limited led by William Orr, and the London-based Mount Elliott Company controlled by London and French shareholders. The Australian representative and General Manager of Mount Elliott was the determined and innovative William H. Corbould. His story is told in *Broken Hill to Mount Isa*, edited and annotated by Ian Hore-Lacy.

With the completion of the railway, Corbould commissioned the Mount Elliott smelters in August 1910: the Hampden Cloncurry smelters followed in May 1911. The First World War years brought prosperity, as the copper price soared from £66 per ton in 1914 to £153 per ton in 1916. Two additional smelters were erected: one by the Corella Copper Company's nearby Rosebud group of mines, and the other by the Mount Cuthbert Company in the northern section of the field.

Despite the harsh living conditions, labour problems, and mine fires, Cloncurry became Australia's premier copper field. The population of the district from Mount Cuthbert in the north to Duchess in the south reached 7,000, with 2,200 at work in the mines and smelters.

The year 1918 saw a record output of 11,600 tons of copper. But 1919 was disastrous, with major equipment breakdowns at Mount Cuthbert, strikes at Mount Elliott, and a fall in the copper price from £112 to £75 per ton in five months. By 1920 the collapse of the field was complete.

Corbould had, since 1911, tried to merge the Hampden and Mount Elliott companies, to enable more balanced and efficient development. With additional capital to develop widely scattered mines and a central treatment plant, he believed Cloncurry could become one of the world's great copper camps. Besides his private venture into Australia's only cobalt mine at Mount Cobalt, he tried in 1921 to develop Mount Oxide, but he was rebuffed by the shareholders and resigned.

Two years later Corbould returned to the district to inspect the new silver-lead-zinc deposits at Mount Isa, which John Campbell Miles had found in February 1923.

W.H. (Jimmy) Corbould, who played a main role in the early development of Mount Isa Mines Ltd.

Discovery and Development of Mount Isa

It is not possible for me to cover the discovery and development of Mount Isa in any detail. Professor Blainey's *Mines in the Spinifex* and the researches and readings of the History Department of James Cook University, under the guidance of Dr Kett Kennedy, deal with these events in detail. I also commend to you the Presidential address of Julius Kruttschnitt, *A Brief History of Mount Isa Mines' Financial Structure; Its Trials and Tribulations*, and the Review Lecture by K. B. Gross, given

Mount Elliott copper mine and treatment works near Cloncurry, 1911.

at the Annual Conference of The Institute some 38 years ago on 18 August 1952 in the old RSL Memorial Hall in Mount Isa.

Douglas MacGillivray (centre), who consolidated the leases at Mount Isa in 1924, with E.C. Saint-Smith, the first mine manager (right), and Bill Purdy, one of the pioneers.

Many names are associated with the early history of Mount Isa: the discoverer John Campbell Miles, Bill Simpson, Con Davidson, Bobby Lawlor, Douglas MacGillivray, Arthur Campbell, E. C. Saint-Smith, Randolph ('the Reckless') Bedford and many others. However, the name of W. H. ('alias Jimmy') Corbould stands out. As K. B. Gross records: 'W. H. Corbould must always be credited with having founded the enterprise, for he correctly appraised the potential of the struggling prospect and foresaw what could be done, despite the handicaps of geography, climate and the need for heavy expenditure'.

Corbould was 'an energetic but retiring man whose single minded pursuit of mining excellence in all its technical and management facets took precedence over all else including family. Doing the right thing for his mine, be it Mount Elliott or Mount Isa, was to him more important than the effects of treading on a few prestigious toes. Mount Isa Mines can perhaps thank him for that'. Corbould was a member of The Institute till his death on 16 March 1949 in the south of France. Both he and his work are commemorated by The Institute in the Student Essay Prize which carries his name.

Corbould, who was introduced to the field by Douglas MacGillivray, the Cloncurry publican and storekeeper, floated Mount Isa Mines Limited in January 1924 and finally consolidated all the leases on the field under the control of the company by November 1925. After efforts to raise capital, including disappointments with Melbourne mining financiers and Anglo American, he finally effected a deal in June 1927 with Leslie Urquhart, Chairman of the Russo-Asiatic Company. Urquhart, a mining entrepreneur who lost his mining empire in Russia in the Bolshevik revolution, was another remarkable man. His contribution to Mount Isa Mines stands out like the headframe of the shaft that bears his name and is summarised by Dr Kennedy in his book *Mining Tsar – The Life and Times of Leslie Urquhart.*

Leslie Urquhart, chairman of Russo-Asiatic Consolidated Limited, saw in Mount Isa an opportunity to make up for the loss of his Russian mining empire during the 1917 Revolution. He financed Mount Isa Mines Ltd, which had such a troubled development that although launched in 1923 did not pay its first dividend until 1947.

Urquhart took up the venture when others turned it down; he found the capital and dictated the pattern of exploitation. It was his breadth of vision which determined that Mount Isa be developed on a large scale, using the latest mining methods and technology. He insisted that it be done by a single company, that intensive drilling be conducted on which long-term programmes could be formulated, and that the best available expertise be employed. Urquhart personally secured the co-operation of the Queensland Government, the principle of freight concessions and the changes to local mining regulations. He created the company town to maintain harmony in the workforce, a concept alien to Australia at the time. He brought to Mount Isa not only the successful aspects of his Siberian experiences but his forceful leadership and enterprising spirit which steered the venture through the financial holocaust, even at the cost of his own fortune, and temporarily his reputation among British investors. His energy and foresight was his legacy.

Urquhart did not see his dream realised. His health was weakened by the strain, and he died of pneumonia on 13 March 1933 in London.

Mount Isa's insatiable appetite for capital was beyond the resources of Urquhart and his companies, and the American Smelting and Refining Company came to the rescue. Later, even the ASARCO Vice President, H. A. Guess, despaired and said that, the decision to invest in Mount Isa was 'the one serious blunder I have made in my long career over thirty years with American Smelting and Refining Company'.

Finally, under the leadership of Julius Kruttschnitt, who arrived at the mine in December 1930, the long battle was won. The first ore was hoisted in May 1931 and the mill and smelter started the following month, but it was not until the financial year ending 30 June 1937 that the Company was able to show a net profit of £70,309. Julius Kruttschnitt in his 1952 Presidential address summarised:

> As you have doubtless observed, it required six financial shots in the arm to bring Mount Isa Mines Limited through the depths of the depression and into the ranks of profitable mining

Julius Kruttschnitt, who played an important role in the establishment and development of Mount Isa Mines Ltd from his arrival in December, 1930. He retired in 1953.

Mount Isa Mines' Urquhart shaft and treatment plant, about 1930.

> ventures . . . Financing a low grade, mass production mineral deposit requires a high degree of faith in the ultimate outcome backed by an unswerving financial policy.

With the discovery of the copper orebodies in the hanging wall of the Black Star and because of the Second World War emergency, the company suspended lead production and produced blister copper for three years between 1943 and 1946, using old smelter equipment that had lain idle in the Cloncurry district for over 20 years. Even before the first dividend was paid, Kruttschnitt reported to shareholders: on 25 January 1945, 'Your directors have under consideration a comprehensive plan for the post war realisation of a dual copper-lead, mining, milling and smelting venture requiring a very substantial expenditure of money for copper milling and smelting plant, and approximately twice as many employees as now engaged'. Copper ore production ceased and lead ore production resumed in 1946.

Pouring copper at Mount Isa.

Development drilling in T62 Decline, the deepest area of the mine at Mount Isa Mines.

Almost seven years later, in January 1953, the simultaneous production of lead bullion and blister copper commenced. The same year Julius Kruttschnitt retired as Chairman and C. R. Hilton, 'Uncle Charlie' as he was affectionately known, also retired as General Manager and returned to the United States; they were succeeded by Sir George Fisher, CMG, and K. B. Gross respectively.

Intensive exploration financed from earnings, which Fisher instigated, resulted in the discovery of the 500, 1100 and 3000 orebodies, together with increased lead reserves at Mount Isa and in the Northern Leases, with the latter developed as the Hilton mine. These additional ore reserves provided the basis of continued expansion. Research and development led to new copper refining technology (the Isa Process), new smelting technology (Isasmelt), and Jameson cell mineral flotation technology. The most fundamental research of all was the vigorous search for new orebodies.

Production from Mount Isa, from the start of production in May 1931 to 8 April 1990 was 130,920,861 tonnes of copper ore, 91,083,740 tonnes of lead ore, 4,590,317 tonnes of crude lead, 11,479,307 kilograms of silver in crude lead, 3,755,209 tonnes of blister/anode copper, 7,044,382 tonnes of zinc concentrate.

In 1990, MIM Holdings Limited, the holding company formed in 1970, was one of Australia's largest mineral based companies and ranked high internationally. The Mount Isa mine and smelters were complemented by copper refining at Townsville, lead refining at Northfleet in the United Kingdom, and zinc refining in Germany, with a range of finished, semi-finished and associated products from these and other locations.

In addition, the Company produced coal and coke in Queensland, gold in Queensland and the Northern Territory, and had a third interest in the Porgera gold mine in Papua New Guinea. It had shareholdings in a number of American, Canadian, German and English companies

C.R. ('Uncle Charlie') Hilton. General Manager at Mount Isa Mines for many years.

Crushing plant and ore conveyor at Mary Kathleen uranium mine, during its second period of operation.

including a 24.5 per cent interest in ASARCO Incorporated, the company which came to its rescue with funds and technical knowledge in the 1930s.

Mary Kathleen Uranium

The uranium boom in the early 1950s resulted in the discovery of the Mary Kathleen uranium deposit in July 1954. A syndicate of eight, including Clem Walton, a Mount Isa taxi driver, and Norman McConachy, an underground timberman, discovered it in a fine piece of prospecting with Geiger counters near the old copper town of Rosebud, where Walton had lived as a boy.

Headframe at Mount Isa's F58 shaft. The shaft was sunk in 1973 to a depth of 1065 metres.

The syndicate sold out to Australasian Oil Exploration Ltd in August 1954 for £250,000 in cash and a 9 per cent interest in the project. The Rio Tinto Company of the United Kingdom later bought a controlling interest (56 per cent). In 1962 The Rio Tinto Mining Company Limited and Consolidated Zinc Proprietary Limited formed Conzinc Riotinto of Australia Limited, which then held 51 per cent. Ore reserves were estimated at 6,600,000 short tons at an average grade of 0.143 per cent uranium oxide. A township was built in 1958 with all modern amenities.

Mary Kathleen's first period of production was from 1958 to 1963, when it sold some 4,100 tonnes of uranium oxide to the United Kingdom

Atomic Energy Commission. It was then on a care-and-maintenance basis for nearly 13 years, after which it produced a further 4,800 tonnes and finally closed in 1982. By 1987, all facilities had been removed and the site rehabilitated, at a total cost of $17,500,000.

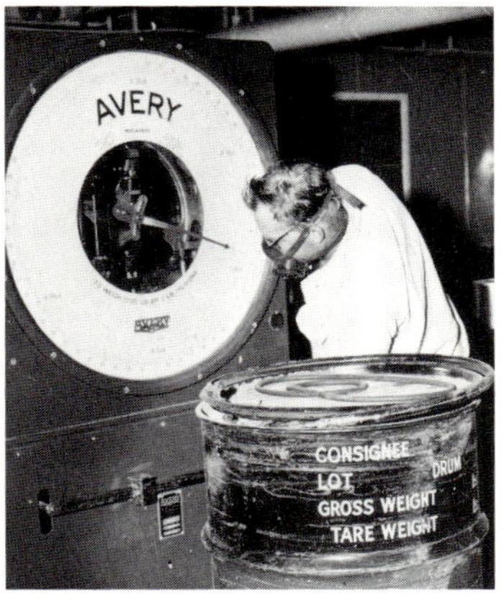

A drum of uranium concentrate about to leave Mary Kathleen, during its first period of operation, from 1958 to 1963.

Other Mining Activities

Other past and current mining activities in the north-western region of Queensland, while small by comparison with Mount Isa, have contributed to the prosperity of the region.

The Starra gold and copper mine, 150 kilometres south east of Mount Isa and 25 kilometres south of the old Mount Elliott mine, was brought into production by open-cut mining in 1987. In its first full year, it produced 85,700 ounces of gold and 9,500 tonnes of copper.

The old Mammoth gold mine was worked by two Afghan gougers Zergoon Shah and Neurallum Shah during the Second World War and later by others. A number of Italians such as Foschi and Angeli had leases

Mount Isa Mines, 1991.

which were purchased and developed by Vam Ltd (VAM). After the collapse of VAM, Gunpowder Copper, a joint venture of Consolidated Goldfields Australia Ltd (CGFA) (70 per cent) and Mitsubishi Development Pty Ltd (30 per cent) treated 1,700,000 tonnes of ore for a production of approximately 42,000 tonnes of copper. In 1990, Adelaide Brighton Cement Holdings Ltd was redeveloping the old mine on a small scale using ore leaching, solvent extraction, and electrowinning. Production was planned to recommence at 7,000 tonnes of copper metal and 2,000 tonnes of copper sulphate a year.

Rail loading facility for phosphate rock at Phosphate Hill.

Broken Hill South and Phosphate Development

Broken Hill South's venture into phosphate in north-west Queensland grew out of its search for a mining project to replace its diminishing asset in its mine at Broken Hill.

By 1966 BH South had identified 10 phosphate deposits in the Mount Isa region, estimated to contain more than 2,000 million tonnes of phosphate rock, with an average grade of 17.5 per cent P_2O_5. With rising phosphate prices in 1973, South announced a project to produce 1,000,000 tonnes of upgraded phosphate rock for domestic use and export per year from Phosphate Hill, near Duchess, under the management of Queensland Phosphate Limited, a wholly owned subsidiary. Open-cut mining began in 1975 when the economic and market climate were highly favourable, but these quickly changed. The Whitlam Government's removal of the superphosphate bounty and refusal to grant viable export permits, plus rising inflation and a changed market, resulted in closure of the project on 30 June 1978 after the expenditure of over $100,000,000.

The project's lack of success led to the decline of BH South Ltd and it was taken over in 1980 by Western Mining Corporation Holdings Ltd. A project to produce phosphoric acid and high analysis fertilizer, using rock from Phosphate Hill and sulphuric acid made from Mount Isa smelter gas, was later designed and investigated. In September 1990 it was announced that the greatly overvalued Australian dollar and the high capital cost did not permit development at that time, but that the project would be kept under review until conditions improved.

Future Outlook

North-west Queensland's history shows it to be one of the great mineral regions of the world. It remains after more than 100 years a very important contributor to Australia's mineral output, and will maintain this position for many years ahead. By 1990, announcements of further base metal discoveries in the region showed it had potential for an even higher level of Australian and world importance. Minerals from north-west Queensland could be expected to continue to serve the community well.

I am indebted to Ted Davies, former Research and Technical Manager at Mount Isa, and Gilbert Ralph, MBE, for their contributions to this paper, which was presented at Mount Isa on 9 October 1990.

Northern Territory

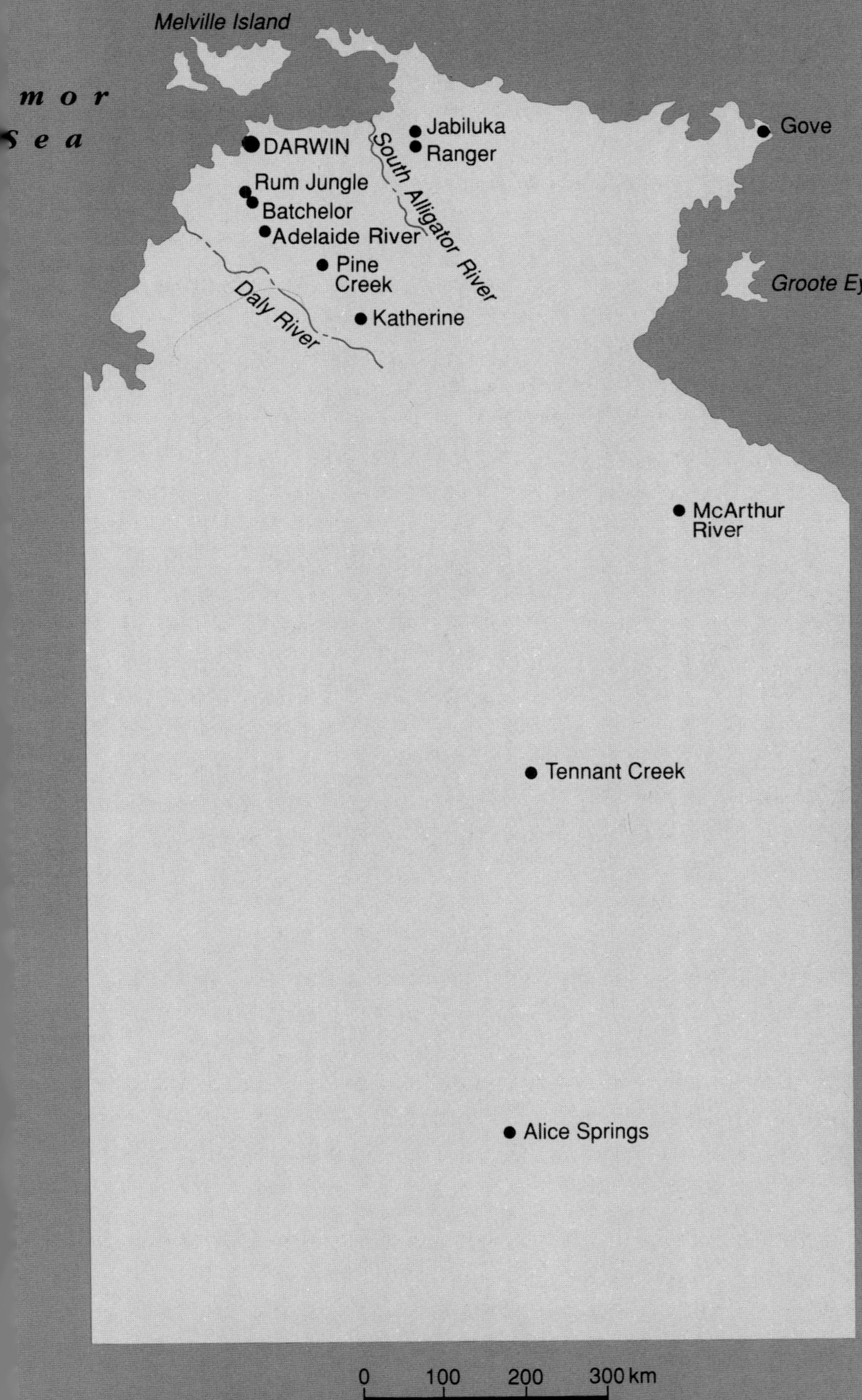

Manganese from Groote Eylandt

Europeans may have settled in Australia early in the 18th century, some 70 years prior to the arrival of the First Fleet. A community of 50 or more people (of Dutch, German and Scandinavian descent) is believed to have lived 500 kilometres north of what is now Perth, apparently settling there after a shipwreck. Just think how different our recent history might have been if European settlement had spread from the north-west rather than the south-east!

As it was, much of northern Australia was largely neglected until after the Second World War. More recently though, greater attention has been focused on this part of the country, and we have come to know its mineral wealth and appreciate the importance of its proximity to the Asia-Pacific region.

Early Days

In the 17th century, the Dutch mapped much of the northern coast of Australia. Abel Tasman gave Groote Eylandt its name during his second journey of exploration, in 1644. In July 1802 the Englishman Matthew Flinders left Sydney in the *Investigator* to explore and chart the northern coastline. By November he had reached the Gulf of Carpentaria. In 1803, he named Groote Eylandt's most prominent feature, Central Hill, and recorded the presence of ironstone and quartz on the island. His observations of the Blue Mud Bay area on the mainland suggested that manganese might occur.

More than a century later, in 1907, the South Australian Government geologist, H. Y. L. Brown, noticed small manganese outcrops, at a time when British steelmakers were looking for an alternative source to the Russian fields. If we take into account the problems of communication and travel in that era (a telephone trunk line had for instance only just opened between Sydney and Melbourne), it should not come as a surprise that no further investigation of the deposits was made.

Giant scraper in the open-cut manganese mine at Groote Eylandt, 1970s.

In the 1920s a mission station was established on Groote Eylandt. Later, during the Second World War, the Royal Australian Air Force had a base here, but this was not kept up after the war. Australia entered the post-war period knowing very little about Groote Eylandt's mineral potential.

Mineral Exploration in the Post-War Period

From the 1950s onwards there was tremendous activity in minerals exploration and development in Australia. Much of the exploration was in the northern part of the continent, catalysed, in the view of Professor Geoffrey Blainey, AO, by the uranium discoveries at Rum Jungle in 1949 and at Mary Kathleen in 1954. In 1955, according to Professor Blainey:

the discovery of bauxite deposits, capable of nourishing a great aluminium industry, suggested that tropical Australia might soon rival southern Australia in mineral wealth.

Interest was not restricted to what were then the new metals, such as aluminium and uranium, but included minerals for which there had long been a market: iron ore, nickel and oil being of note.

That same year geologists from the Frome-Broken Hill Company (owned mainly by large international mining and petroleum companies and not connected with BHP) visited Groote Eylandt and sent samples of manganese ore to their Melbourne office. But, as the company's main interest was oil, they did not proceed further.

In November 1960, government geologists, engaged in a regional geological survey, noted the outcrops of high grade manganese and took samples, which were found to contain 50-55 per cent manganese. The following year, one of the geologists did limited test-pitting and an inspection of other promising areas. The work indicated the presence of economically-valuable deposits, 'put[ting] an end to yet another deficiency in Australia's mineral economy'. The Church Missionary Society (CMS), took out a permit on behalf of the local Aborigines, entitling it to prospect and seek mining titles to an area of 530 square kilometres.

Back hoe excavation for removing over burden at Groote Eylandt.

Imports of manganese ore for metallurgical use were at that stage costing Australia about £500,000 a year. These imports were supplemented with small amounts of domestically-produced ore, which came from a number of small, hand-worked deposits in central and Western Australia, and along the Great Dividing Range. Imports of ferro-manganese had also been steadily increasing and by 1961 were valued at more than £2,000,000. This import bill dropped markedly with the commissioning in March 1962 of a ferro-manganese plant by the BHP

Drilling to establish the mining plan, Groote Eylandt, 1964.

Concentrator plant, Groote Eylandt, 1974.

subsidiary, Tasmanian Electro-Metallurgical Company (TEMCO), using imported ore, at Bell Bay in Tasmania.

While the government geologists' investigations were under way, BHP became interested in the manganese deposits on Groote Eylandt. In 1962 a company representative, Bill Smith, made a one-day visit to the island and reported back to management. Later that year some preliminary testing was done and BHP began negotiations with the CMS, leading to agreement on 7 June 1963. In October 1963 BHP's Chief General Manager, Sir Ian McLennan, KCMG, KBE, announced that mining would proceed and the Groote Eylandt Mining Company (GEMCO) was formed in 1964. Extensive exploration lasted from 1963 to 1967.

Preparing to blast at Groote Eylandt manganese mine, 1966. Each blast shifted thousands of tonnes of manganese ore ready for quarrying.

Development in the Early 1960s

Many of the challenges faced by those charged with the development of the mine on Groote Eylandt were common to miners in Australia at the time. Working in a remote location always brings hardships. The first camp, a prospecting camp, was on the Emerald River. Later everyone lived in tents where the town of Alyangula was later developed.

Supplies were brought in on the *Alagna* which sailed weekly from Brisbane. This gave rise to the saying, 'It's on the next Alagna', meaning the speaker had not yet done whatever he had been asked about. A plane visited once a week. The airport coach was a truck. In those days it was too expensive to transport cars to the island.

The island had only one radio phone, which employees could use for personal calls only on Saturday mornings. Since then, life in remote areas has been transformed by advances in communications and transportation. It is now hard to imagine the isolation experienced by those who developed the mine in the mid-1960s.

Relations with the Local Population

Although many of the early challenges were like those in other parts of Australia at the time, an important difference on Groote Eylandt was that the mining was on Aboriginal land. And, just as many of the communications devices we rely upon today did not then exist, neither did the codes of conduct and land rights understandings to which we are now accustomed. The various parties had to agree among themselves.

The agreement signed on 7 June 1963 included measures relating to the employment of Aborigines and the conduct of GEMCO employees toward the local people. Since then, relations between the Company and the local people have steadily built upon a foundation of mutual understanding and mutual respect, and an important aspect of the operations has been the willingness of all managers to continue that understanding and respect. It is important to remember that these relationships are between individuals and depend upon individuals' integrity.

Steady Development

Construction of the port, township and treatment plant went ahead following the formation of GEMCO in 1964. The first shipment of ore left Alyangula on 17 March 1966, bound for Bell Bay, there to be turned into ferro-manganese and silico-manganese for use in BHP steelworks. In September 1966 the first export shipment of manganese ore left Groote Eylandt for Japan. The export market was later diversified to include Taiwan, South Korea, the People's Republic of China, the Soviet Union, the United States and several European countries. Japan became

Classifiers in the concentrator plant at Groote Eylandt, 1971.

GEMCO's largest customer. In 1966 mining production capacity stood at 200,000 tonnes of ore per annum. This increased rapidly and by the end of the 1960s was about 500,000 tonnes per annum.

Late in 1968, preparations began for design and construction of a beneficiation plant to produce 1,000,000 tonnes of lump ore a year. This would enable GEMCO to mine a wider range of ores, increase its range of products, and ensure consistent product grades. The plant was commissioned by early 1972, as were extensions to the township and port facilities. The concentrator began full operation on 20 March 1972. Production soon reached 1,000,000 tonnes a year, and doubled again by 1976. In the first decade of operation, 9,000,000 tonnes of manganese ore were produced, of which 7,400,000 tonnes was exported.

After minor upgrades in 1975 and 1979, a Heavy Media Fines Plant was added in 1982 and the Heavy Media Drum Plant converted to automatic control. Further automation was made possible after another major upgrade in 1986. This also saw the concentrator's annual capacity increased to about 2,400,000 tonnes.

Increases in productivity do not come about from plant improvements alone. Production records were also broken through the determination of people to meet their goals. GEMCO grew from being mainly a supplier

In the township of Alyangula, Groote Eylandt, 1974.

Loading manganese at Groote Eylandt, 1972.

to BHP steelmakers to a high-quality producer selling into international markets. Annual production settled down at around 2,000,000 tonnes of ore, with about 80 per cent going to the export trade, comprising between 10 and 15 per cent of world trade in manganese ore. By 1990, GEMCO had shipped well over 30,000,000 tonnes of manganese ore from Groote Eylandt.

Rehabilitation

Land at Groote Eylandt is systematically restored after mining, with all efforts directed to creating self-sustaining ecosystems like those that occur naturally near the mining areas. About 30 hectares of mined land are back-filled, covered with top soil, and sown each year. Seed collection of about 25 native species is done on contract by local Aborigines. This seed is processed and stored, then broadcast over areas to be restored. Regular surveys of flora and fauna follow restoration.

Future Outlook

In 1990, the outlook for manganese was good. Demand was strong in the ore and sinter markets and the trend was expected to continue as Eastern European countries adopted a rational economic approach,

Sir Arvi Parbo, (2nd from right), during a visit to Groote Eylandt in 1990, with (l-r) Bob Wunder, Manager GEMCO, Lady Parbo, Jimmy Yantarnga (President of Angurugu), Jabina Lalara and Jambana Lalara (Former President of Angurugu).

closed unprofitable mines and sought higher-quality, lower-cost imported ores.

As mentioned, good relationships between miners and the local Aboriginal population have been built up steadily over many years. It is vital to both, the miners and the locals, that these good relationships are maintained and built on. Some years ago one of the local elders observed:

> White people have come here because of the discovery of the manganese, from many different places, from all over Australia and from overseas. Young people find this easier to understand than the old people. Each year brings more changes, and for some people all this change has come too fast, especially for the old people. I think that we can keep the best of both ways.

Since then, the years have brought many changes on Groote Eylandt, with the pressures on traditional Aboriginal values being particularly great. The desire to see the local community 'keep the best of both ways' amidst all this change remains a challenge for all on Groote Eylandt.

I am grateful to Vicki Lukins, for her major contribution and Alan Wright and Warren Paull of BHP Minerals for their assistance in the preparation of this paper, which was presented at Groote Eylandt on 10 October 1990.

Gold and Uranium in the Northern Territory

Mining and minerals were interwoven into the fabric of Northern Territory development from its inception.

The first recorded discovery of minerals in the 'Top End' was the gold Frederick H. Litchfield found in the Finness River in 1865. Gold's presence there was confirmed in 1869 by George W. Goyder, who was surveying the route for the Overland Telegraph Line. The Northern Territory Gold Prospecting Association, formed in Adelaide in 1871, sent a well-equipped prospecting party, which arrived in Darwin by ship on 11 March 1872. It headed for Pine Creek, found payable gold there, and started a minor rush.

In 1873-75 many South Australians and Victorians, spurred on by the prosperity of the rich Victorian fields, invested heavily in Northern Territory companies and syndicates. By May 1874, sixty companies had been floated on the Adelaide Exchange to develop mines in Northern Territory, many of them based on exaggerated claims of rich reefs on leases which had hardly been explored.

Lack of transport and roads, and the six months wet season, disrupted all attempts to develop mines. Their remoteness from Adelaide, the source of most supplies, added substantially to costs. The construction of the Overland Telegraph Line in 1871-72 provided rough tracks and numerous watering points, but beyond the Line the virgin country proved formidable.

The telegraph encouraged brevity and the use of superlatives in describing new discoveries. In 1874 Mining Warden Conner remarked on 'the barefaced false reports telegraphed to Adelaide by interested parties as to the value of claims and the quantity of stone available for crushing'. While he was familiar with all the tricks of the promoters elsewhere, 'these most decidedly beat them out of the field'. Information published by companies was not as well scrutinised as it is today. One report of the time stated that 'the assays from sampling [were] so confidential that they [could] not be disclosed', another that 'in the leases there [was] a 60 mile ridge of solid gold'.

The Union Gold Mining Company, reputed to have won 800 ounces of gold from 10 tons of ore, paid the Territory's first dividend in 1875.

The climate and remoteness of northern Australia added to the problems of attracting labour to the fields and opened the way for recruitment of Chinese from Singapore and Hong Kong on two year indentures. By 1879 there were an estimated 3,400 Chinese, seven times the number of Europeans, in the Territory and many more followed. During the building of the railway from Darwin to Pine Creek in 1887-89 nearly 3,000 Chinese came as labourers, bringing the total to 6,200.

Chinese labourers came to the Northern Territory to help build the Darwin to Pine Creek railway, to work as miners and at other tasks.

Many of the Chinese who came to work on the railways and in the mines turned to other occupations in Darwin.

Despite the many difficulties, the Territory's aggregate gold production to 1889 was estimated at 256,690 ounces. For the next 40 years prospectors ranged widely over the Territory and by 1930 gold had been won from the MacDonnell Ranges, at the Tanami, Tennant Creek and Housechilds rushes, culminating in extensive reef mining at Pine Creek, Union Reefs, Eureka, Howley, Zapopan, Yam Creek, and many other places.

Tin had been mined at Mount Wells, Mount Shoobridge and Maranboy; copper near the Daly River, Mount Diamond, Coronet Hill and Mount Elliston; wolfram at Hatches Creek, Yeuralba and Wolfram Hill; and lead, zinc and silver at the rich Evaleen Mine, Silver Spray, and several other localities.

These were also the boom years of English speculation when Horatio Bottomley through a variety of companies developed or controlled most of the mines. Lack of capital (most money raised went into promoters pockets), shortage of skilled labour, lack of infrastructure, seasonal conditions, and transport delays precluded effective development. Without the railway from Darwin to Pine Creek, which opened in 1889, to Katherine in 1917, and from Adelaide to Alice Springs, which was completed in 1929, much less would have been achieved.

During the early 1930s there was a resurgence of interest in gold mining, the price having doubled when England abandoned the gold standard and the Australian pound was devalued by 25 per cent. The gold price rose from £4.5.00 per ounce in 1929 to £8.10.6 per ounce in 1934. Sir Lindesay Clark in his book *Built on Gold* records how:

> In 1932, C. G. Gibson and I inspected extensive leases held at The Granites in the anthill and spinifex country about 400 miles north-west of Alice Springs. It was a gruelling trip and terribly hot... For all our efforts we saw no deposits that warranted exploration. We were also rather uncertain of the Aborigines in the area. At one time when a group came shouting and roaring at us it was a relief to realise that all they wanted was bully beef of which, luckily, we had a plentiful supply.

Lasseter

Harold Lasseter became a legend because of a rich reef he supposedly discovered near the Petermann Ranges, in the south-west corner of the Northern Territory. During a return expedition to locate the find Lasseter became separated from the party and died in the desert.

The People

Prospectors and their families who bore the hardship and isolation produced the mineral wealth of the Territory from a harsh land of two faces. 'The Wet' and the 'The Dry' dominated movement, production and efficiency. Only mining centres close to the Northern Territory railway could effectively operate throughout both seasons.

In the early years alluvial mining and crushing batteries virtually stopped because of the lack of water during the 'The Dry'. Most miners then returned to Darwin for sustenance until wet season rains again provided water for processing.

W. G. Masterton found and worked mineral discoveries in the Jabiluka area with the aid of Aboriginals. He devoted his life to them until his death in the 1970s, aged more than 90 years. He named the various occurrences after his helpers: Black Charlie and Yellow Girl were two of them. The 30 per cent copper ore was hand picked by the Aboriginals, and with the proceeds Masterton bought food and clothing for them. He lived in a cave below the escarpment not far from the Redbank Mine.

Lindesay Clark (left), C. G. Gibson (centre) and a Mr Wetherall at The Granites in 1932.

Death certificate of Harold Lasseter, one of the best-known Northern Territory prospectors. He claimed to have found a rich reef near the Petermann Ranges, but became separated from the return expedition to locate the find and died of starvation. Nobody else has ever been able to find the reef.

NORTHERN TERRITORY OF AUSTRALIA

Nº 7807

Registration of Births, Deaths and Marriages Ordinance 1973

COPY OF ENTRY NO. 28/OCA **IN REGISTER OF DEATHS**

Personal Particulars of Deceased:	
1. (a) Full name	Henry. Lewis Bell Lasseter
(b) Occupation	Prospector
(c) Usual Place of Residence	Orient Road Kogarah Sydney N.S.W.
2. Date and Place of Death	January 30th 1931 Shaw Creek Peterman Ranges C. A.
3. Sex and age	M 50
4. (a) Place of Birth	Melbourne Victoria
(b) Length of Residence in Australia	Life
5. Conjugal Status	
6. (a) Place of Marriage	
(b) Age at Marriage	43 years
(c) To whom Married	
(d) Names and ages of living issue in order of birth	1 Male. 2 Females.
(e) Number of issue dead	Males; Females
Medical Particulars:	
7. Cause of Death	Starvation
8. Duration of Fatal Illness	

His double bed suspended from the roof of the cave can be seen today. He was buried beside the mine.

Transport and Communications

Away from the railway, people throughout the Territory depended for transport largely on bush tracks pioneered by the pastoralists and the miners. The lack of suitable transport routes and communications was a deterrent to development until the Second World War. The bombing of Darwin in 1942 prompted the Federal Government to build narrow sealed roads to the railhead at Alice Springs and from just north of Tennant Creek to the railhead at Mount Isa. For the first time, the Territory had all-weather transport routes to other parts of Australia and a main north-south road.

Examining core samples, Ranger uranium mine.

Uranium

Despite the improved infrastructure, the Northern Territory's mining industry languished after the Second World War. Darwin was a bombed-out shell with fewer than 5,000 people. The way the war ended, after the use of atomic bombs on Hiroshima and Nagasaki in 1945, created a demand for uranium for weapons, and it was uranium that created the next wave of activity. The Federal Government, through the Bureau of Mineral Resources, began a search for uranium in Australia, offering a $50,000 reward for discovery of an economic deposit.

Prospector Jack White found uranium at Rum Jungle in 1949, thus focussing the Government's attention on that area and the Northern Territory generally. The deposit was developed under the management of Territory Enterprises Pty Ltd, a subsidiary of Consolidated Zinc (now

CRA). Prime Minister Rt Hon Sir Robert Menzies, KT, (PC), CH, officially opened the Rum Jungle plant in 1954, and mining continued until 1963. The town of Batchelor, which still thrives today, was created for the work-force, many of whom came from elsewhere to the Northern Territory.

Lack of success in finding more uranium deposits prompted amendments to the Atomic Energy Act to allow private industry to search for, develop, and export uranium under Australian Atomic Energy Commission control. A uranium boom followed, many new Australian companies were formed in 1953-54, and, in competition with the larger and longer-established companies, set about the task.

In 1953, Bruce Walpole, a Bureau of Mineral Resources geologist, was shown a copper-bearing structure in the South Alligator River valley by Jim Callinan, the Gimbat pastoral property lessee. Walpole discovered the radioactive outcrop further up the hill from the copper-bearing lode and named it Coronation Hill in recognition of Queen Elizabeth's Coronation on that day. Several companies explored the South Alligator River valley and made a number of discoveries. The boom became a bust in 1955 with an over-supply of uranium from new discoveries in Canada and the United States. Contracts were eventually put in place extending to 1964, but exploration around the escarpment was suspended.

This short-lived exploration boom, followed by a decade of production, was the catalyst that encouraged Darwin to grow from a town of 5,000 without much hope to a small modern city of 17,000 in less than eight years. It also focussed exploration by major companies on the Northern Territory and the renewed market for uranium for electricity generation

Dyson open-cut mine on the uranium field at Rum Jungle.

The treatment plant at Ranger uranium mine. Ranger makes its own sulphuric acid, which it uses to leach uranium from crushed ore.

in the late 1960s led to significant new discoveries, such as Queensland Mines' Nabarlek deposit in 1970, Peko-Wallsend/EZ Industries' Ranger 1 deposit in 1970, Noranda Australia's Koongarra deposit in 1971, and Pancontinental Mining's Jabiluka deposit in 1972.

Alumina

A wartime air base on the Gove Peninsula had runways of reddish local material. Not until 1952 was it recognised that the material was bauxite of economic grade. Its remoteness and the location of the deposit in an Aboriginal Reserve devoid of developed roads and ports contributed to difficulties of exploration and development. Gove Aluminium Ltd (30 per cent) and Swiss Aluminium Australia Ltd (70 per cent) were able to bring the deposit into production in 1974. It has since expanded to become a significant contributor to the prosperity of the Territory.

Petroleum

Natural gas was discovered at Mereenie in 1964 and at Palm Valley in 1965, both in the Amadeus Basin, west of Alice Springs. The

Floating oil production, storage and offloading facility, Challis Venture, Timor Sea.

subsequent development of the field for oil and gas production and the building of a gas pipeline to Darwin have been important to the Territory.

Discovery of oil in the Timor Sea in the early 1980s resulted in the Timor Sea becoming Australia's second largest oil province, after Bass Strait, and one which will be of significant value to Australia.

Value of Mineral Production

Until 1905, gold contributed more than 80 per cent of the total value of the Territory's mineral production but then it lost its position to copper, tin, and wolfram. Alarmed at the declining production, the South Australian Government (which then administered the Territory) in 1904 embarked on testing the various reefs by diamond drilling, but this was far from successful. Recorded gold production of the Northern Territory peaked at 26,354 ounces in 1881, after which it fluctuated between 10,000 and 25,000 ounces annually until 1906. It then steadily declined until the Tennant Creek field was discovered and developed in the 1930s.

Post war redevelopment and expansion saw a substantial increase in gold production, with highs of 258,234 ounces in 1972 and 548,000 ounces in 1989. Total production from the Northern Territory to 1989 amounted to over 5,800,000 ounces. In 1988 the value of mineral production in the Territory was $1.2 billion, of which bauxite and alumina accounted for 28 per cent, uranium 21 per cent, oil and natural gas 19 per cent, and gold 16 per cent.

Alumina refinery at Gove. Bauxite from the mine travels by conveyor belt from the open-cut mines to the refinery.

Government

South Australia was early to realise its inability to make a success of Territory development and eventually prevailed on the Commonwealth Government to take it over. In 1911, the Commonwealth reluctantly accepted. Because of the exodus of Chinese, the non-Aboriginal population had by then declined to 2,846, of whom 1,182 were European. During the First World War, when metal prices were high, there was a revival of production but the sharp fall in metal prices in 1921 caused another decline.

Many Parliamentary Inquiries on Northern Development followed the Second World War, several recommending a port site at the mouth of the McArthur River, a railway connection with Mount Isa, and completion of the North-South Railway. None of these seemed justified to the Canberra politicians and only single-lane sealed roads, for the beef industry, were built.

The subsequent history of government involvement in the fortunes of the mining industry is well known. Until self-government, in 1978, the industry was very much at the mercy of distant bureaucrats and to a large extent still is, especially with respect to pricing of uranium, export control of all minerals, and various environmental and Aboriginal matters. The Northern Territory Department of Mines Annual Report for 1989 said:

> Australia continues to block mining operations for marginal environmental reasons despite balance of trade problems the country is experiencing.

We desperately need a rational approach to the issue, and a system which quantifies the overall economic benefit to the country of such projects.

The environmental aspects should be taken into account in this system of checks and balances, and assigned their overall worth in the scheme of things.

This is necessary because, in the present state of play, environmental concerns appear to outweigh even the national interest.

Conveyer belt from the open-cut bauxite mines to the alumina refinery, Gove.

Future Outlook

Despite the restrictions of government policies, the shrinking land area available for exploration, and imposition of the company tax on gold in 1991, the Territory remained the largest per capita mineral producer in Australia. Providing that Federal and Northern Territory Government policies did not further reduce the land available, intensive exploration had the potential to expand mineral production.

Waiting in the wings for clearance for development and production were several major orebodies: uranium and gold at Jabiluka and Koongarra; silver-lead-zinc (reputed to be the largest high grade zinc deposit in the world) at McArthur River; gold-platinum-palladium at Coronation Hill; and gold at Mount Todd. (In 1991, the Federal Government decided against mining at Coronation Hill).

W. J. (Joe) Fisher, AM, President of The Institute in 1984, when the Annual Conference was held in the Northern Territory for the first time.

The Institute in the Northern Territory

Several members of The Institute have gained distinction for their work in the Northern Territory, including Sir John Proud of Peko-Wallsend for his contribution to development of mines in the Tennant Creek area and Keith A. Rowell for his work at Groote Eylandt. They were awarded Institute Medals in 1974 and 1975 respectively.

Clem L. Knight received the AusIMM President's Award in 1984 for his contribution to the discovery of the McArthur River lead-zinc deposit. W. J. (Joe) Fisher, AM, a Councillor of long standing, served as President in 1984 when the Annual Conference was held in the Northern Territory for the first time.

I am indebted to W. J. (Joe) Fisher, AM, Mining Consultant, and Gilbert Ralph, MBE, for their invaluable assistance in the preparation of this paper, which was presented at Jabiluka on 9 June 1990.

Gold and Copper from Tennant Creek

John McDouall Stuart, on his first courageous but unsuccessful attempt to cross Australia from south to north, in 1860, named a slow-flowing waterway Tennant Creek, in memory of his friend John Tennant. Stuart also named Mount Samuel, a feature some six kilometres south-west of the future town, and noted that its pure ironstone top attracted the compass 160 degrees.

The early development of Tennant Creek was associated with the Overland Telegraph Line, the communications link that helped bring Australia into the modern world. The Line was completed in 1872, with a repeater station some 11 kilometres north of where the township now stands. A Government Reserve was declared around the station in 1874, and the site then boasted two stone buildings which still survive.

In 1895 H. Y. L. Brown, the South Australian Government Geologist, panned wash from Bishop's Creek and found traces of gold. Government Geologist Davidson in 1905 extensively sampled numerous quartz reefs but was not impressed, reporting gold at Kurrundi but none at Tennant Creek. Between 1920 and 1930 prospectors casually examined the area, and in 1926 Charles Windley found some gold at what came to be known as the Great Northern. On the same day Jim Ambrose found some green copper ore at the Shamrock, but it was of no interest at the time.

The prospectors had followed the practice in other goldfields in Australia and mainly prospected the quartz ridges, looking for the rich quartz reefs typical of Bendigo and Ballarat, unaware that the gold at Tennant Creek was associated with the lenticular, ironstone bodies.

In time the prospectors began to examine the quartz hematite bodies, and in 1933 some six ounces of gold to the ton was discovered in an abandoned shaft about nine kilometres south-south-west of the Old Telegraph Station, close to the original Great Northern discovery. Nine tons of ore, sent 1,340 kilometres to the Peterborough battery in South Australia, yielded 25 ounces of gold by amalgamation, with 25 pennyweights in the tailings. By August 1933 a rush had set in, the miners closely followed by speculators and investors. Warden Bell and Constable Muldoon opened the first warden's office and police station at the Old Telegraph Station in 1933. When Tennant Creek was gazetted a gold field on 21 September 1935, about 200 men were prospecting and gouging on the field.

They attracted budding merchants and traders, who needed somewhere to establish their businesses. The Old Telegraph Station was on a government reserve and private enterprise was not allowed. Joe Kilgariff, uncle of a future Northern Territory Senator, Bernie Kilgariff, wanted a store, so he built it 11 kilometres to the south, where a supply truck carrying beer had bogged down, possibly early in 1932. The north-south

John McDouall Stuart and his party unfurl the British flag in central Australia. Stuart named Tennant Creek (after his friend John Tennant, an early settler in the Port Lincoln district of South Australia) on his first attempt to cross Australia, in 1860. Stuart noted that the top of Mount Samuel was a mass of ironstone, but not that it had a high gold content.

Completion of the Overland Telegraph Line in 1872 provided better communications for those interested in mining, including speculators who used it to transmit exaggerated claims.

track became Patterson Street and the Old Telegraph Station became known as the Seven Mile. Settlers began to straggle in from the Seven Mile, and a post office was built in 1934.

The one and only alluvial area was pegged in 1937 near the Orlando mine, and worked by individuals for many years, with very modest production. Its main claim to fame was the alleged discovery of a 32 ounce nugget, the largest ever found in the field.

Early Development

One of the first batteries was set up 11 kilometres east of the town, near what would in time be the site of the Peko mine. This was the Empire Mill of the Central Milling Company and the foundations of an old Chilean mill could still be seen in 1990. There had been a four-stamp battery on the Great Northern, but this had no official recognition. Honours for the first official battery go to Fazal Dean, Afghan trader and operator of camel trains, who built 'the Ghan's' battery some three kilometres south of the newly-established town. Three government batteries followed, in various parts of the field.

Production from the fields rose from 62 tons of ore in 1934 to 24,732 tons in 1942. Up to 1952 some 246,974 tons at a head grade of 21 pennyweights per ton had been mined, mostly from small prospects over

2,500 square miles. Before the Second World War 113 mines were active, but during the war only the Eldorado operated, with the Army using its workshop.

From 1946 to 1950 there were 25 mines, but in 1951 this was down to four producing mines: Nobles Nob, Eldorado, Peko, and North Star. About 80 per cent of the gold came from Nobles Nob. Two conflicting economic factors influenced this decline in gold production: the Australian currency was devalued in 1949, increasing the price of gold from £10.15.4 to £15.9.9 per ounce; then in 1951 there was an exodus to Hatches Creek to the south because of high prices for tin and tungsten.

By then, the time of the prospector was passing. The key to unlocking the field was the magnetic susceptibility of the host rock. As John McDouall Stuart had noted, the top of Mount Samuel was a mass of ironstone. Much credit must go to the cooperative efforts of the respective governments that led to several northern surveys including the Aerial, Geological and Geophysical Survey of Northern Australia which began in 1934.

By 1953 the North Star mine was operating in a spasmodic way. It had underground workings with an open-cut being developed and its own treatment plant, but it ceased operations when it ran out of ore. Nobles Nob was operating as a public company, Australian Development N.L., with underground grades of three ounces to the ton or more, and its own mill. Eldorado Gold Mine NL, also a public company, was producing regularly, with an underground mine and a mill, and retreating tailings; it mined grades of up to 60 ounces per ton during its best times, but it closed in 1959. Peko was not producing, having operated earlier as a rather unsuccessful gold mine. Nobles Nob, Eldorado, and the North Star were shallow mines, in the oxidised zone, no deeper than 100 metres or so. The Peko Mine, however, and its offshoot mines of Orlando, Ivanhoe, Juno, Gecko, Warrego and Argo, penetrated into the primary zone where the massive magnetite host rock held the copper, gold and bismuth mineralisation.

Like prospectors everywhere, those in the Northern Territory moved from place to place, carrying their prospecting gear as well as their swag.

Nobles Nob

Jack Noble and William Weaber discovered the Nobles Nob orebody in 1933. Noble had sight in only one eye and Weaber was completely blind, but these two men found this veritable Fort Knox, which yielded 33 tons of gold from 2,000,000 tons of ore until mine closure in 1985.

Noble would find a likely-looking specimen, describe it to Weaber, who would handle it for some time, 'seeing' it with his fingers. The two men would discuss it, pool their conclusions and decide. On this basis, they pegged four adjoining sites: Rising Sun, Nobles Nob, Weabers Find, and Kimberley Kid. They worked Rising Sun first, then Nobles Nob, where the first shaft was sunk in 1934. Jack Noble sold his interest in the leases for £10,000 and bought the Tennant Creek Hotel for £5,000, another gold mine of sorts, but he later fell on hard times and died in 1966 at the age of 80 in the Tennant Creek Old Timers' Home.

Australian Development N.L. at Nobles Nob can claim early experience in the art of rock mechanics. Under the direction of Professor E. L. J.

Abandoned Eldorado mine at Tennant Creek. Of 113 mines active at Tennant Creek before the Second World War, it was the only one to operate during the war, partly because the Army used its workshop.

Potts of Britain's Durham University, they set up a rock stress monitoring programme in 1960, with the objective of mining pillars without the introduction of fill. Disaster struck in August 1967, when some shallow stope arches failed, the winder house and several buildings slipped into the resulting cavities, and the main shaft became inoperable. Fortunately nobody was injured, as the failure happened late at night and the instrumentation had given some warning. The Company switched to open-cut mining, which was a success to the end of the life of the mine in January 1985, after 37 years of continuous operation, with milling of the stockpiles continuing to 1988.

Australian Development N.L., which Poseidon Ltd later acquired, moved on to further success with the White Devil mine, a small mine with above-average grades. It also mined minor tonnages from the Black Angel, Golden Kangaroo, and Golden Forty mines.

Abandoned North Star mine at Tennant Creek. Like Nobles Nob and Eldorado, it was a shallow mine in the oxidised zone.

Peko Mines

The history of Peko Mines N.L. is a fascinating story, from its discovery by prospectors, through unsuccessful operation as a small gold mine, to its graduation into copper below the oxidised zone, the discovery and development of six other mines, a courageous attempt to smelt copper, and the successful unravelling of the metallurgy of bismuth in the Tennant Creek environment.

In 1933 Joe Kaczinski, a Polish prospector, together with Bill Bohning, operating out of the tent town around the Telegraph Station, pegged the Peko leases. The story is that Peko was the name of Kaczinski's dog, which had been named after Pekoe tea.

In 1936 the Aerial, Geological and Geophysical Survey of Northern Australia disclosed a major magnetic anomaly on the Peko lease. Shallow drilling indicated low grade gold. In the 16 years from 1935 to 1951, over 6,000 tons of ore, which averaged about 14 pennyweights to the ton, were extracted.

Peko (Tennant Creek) Gold Mines N.L. was registered in Sydney in 1949, and began operations in a modest way in 1950. During 1950-51 a small but rich dome of oxidised copper was encountered on the 200 feet level in the search for gold, this material being sent to Port Kembla as ore (there were substantial lumps of native copper) for smelting. Also, at this time, the Bureau of Mineral Resources intersected primary chalcopyrite at about 370 vertical feet with a diamond drill hole.

During 1952 and the early part of 1953, the company worked on, in a cautious way, trying to assess the importance of the copper deposit. A high grade secondary enrichment zone was opened up on the 300 feet level, and it soon became apparent that a copper orebody of some importance had been found. The mill started up in 1954, treating 50,000 tons per year of 7 to 8 per cent copper ore.

The Peko mine changed the nature of mining in the Tennant Creek field. It used modern mining equipment and methods, deep for the area (about 380 metres), and was forward-looking and aggressive in its exploration of the mine and the field. The mine produced continuously between 1954 and 1975, yielding about 115,000 tonnes of copper and 315,000 ounces of gold in concentrates, most of it exported to Japan.

The Peko mine also experimented in rock mechanics. Pillar stitching and hanging wall support was introduced in 1964, using 29 millimetre diameter winder rope grouted into drill holes. Underground exploration continued during the life of the mine, locating nine satellite orebodies near the main orebody.

Nobles Nob mine gave up 33 tons of gold between its discovery in 1933 and its closure in 1985. It was a shallow mine in the oxidised zone.

Orlando

Exploration in the Tennant Creek field away from Peko also had almost continuous success. The Orlando mine was brought into production in 1962, at about 30,000 tonnes of gold-copper ore per year. Then came the Ivanhoe mine in 1965, which built up to 50,000 tonnes of copper-gold ore per year. These two mines helped underpin Peko's long term contracts to Japan, and gave valuable experience in the operation of small mines. As satellite mines they had a minimum of personnel, and all ore was treated at the central mill at Peko. Together they contributed about 12,000 tonnes of copper and 145,000 ounces of gold in concentrates.

Juno mine, Tennant Creek.

Juno

In 1967 one of the most remarkable small mines in Australia came into production. This was the Juno mine, just six kilometres from Peko, with the main shaft right in the middle of the road leading from Nobles Nob to the Eldorado mine. This orebody, with no surface outcrop, was a triumph for geophysical exploration and the high skill of interpreting magnetic anomalies by geophysical consultant L. Richardson. Ore reserves were about 455,000 tonnes assaying 59 grams per tonne of gold, 0.7 per cent bismuth and 0.37 per cent copper, in excellent mining ground between about 180 metres and 350 metres below the surface. Juno yielded about 580,000 ounces of gold in bullion and a further 35,000 ounces in concentrates, a total of about 26 tonnes of gold over eight years. It also produced some 2,800 tonnes of bismuth in concentrates.

Bismuth created extra wealth, but also major metallurgical problems, as world custom smelters of copper disliked bismuth in copper concentrate. While the Sumitomo Company in Japan had successfully handled the low bismuth content of Peko concentrates at their small specialised smelter at Kunitomi, this smelter was to be closed down. At the same time the bismuth content of Tennant Creek ores was rising. This emerging bismuth problem led to development of a differential flotation technique, producing a copper-gold concentrate and a separate bismuth-gold concentrate. This later led to planning for a copper smelter at Tennant Creek in which special techniques would be used to control the amount of bismuth in blister copper, and to collect the bismuth as an economic by-product.

Warrego

Another very significant find was the Warrego mine, again using magnetic anomalies to drill for an orebody beneath the featureless surface. Work began in 1966, culminating in the start-up of a 500,000 tonnes a year concentrator in 1972. Before this, the Peko concentrator had been handling ore from five mines, and its experience was used to design a very effective concentrator at Warrego, including a sulphate roaster for final separation between copper and bismuth. The roaster treated bismuth concentrates and bismuth-rich smelter flue dust. Bismuth grades at Warrego were high, averaging of about 0.3 per cent.

Road Transport

Efficient road haulage began with the cartage of Orlando ore for 40 kilometres to the Peko mill in 1962. This depended on a suitable but not costly haul road, good loading and discharge facilities, and a road train that could carry 80 tonnes in side-tipping trailers. This system was still working in 1990.

Smelting

As the original Peko mine and its satellites Orlando, Ivanhoe and Juno were phased out, the Warrego mine and Gecko mine (discovered in 1968, with underground development from 1973) became the focus of operations, with plans for a central concentrator at Warrego and a copper smelter between Gecko and Warrego.

Tests at Outokumpu Oy in Finland gave satisfactory production of matte and the removal of 95 per cent of the bismuth as flue dust using the flash smelting technique. Further tests were done in Sweden on removing the residual bismuth in the matte in a top-blown rotary converter, as it was believed that the bismuth could not be removed in conventional, bottom-blown Pierce-Smith converters. Construction of a flash smelter with top blown rotary converters began towards the end of 1970, with the first blister being produced in December 1973.

Peko mine, Tennant Creek. It penetrated into the primary zone, where the massive magnetite host rock held the copper, gold and bismuth mineralisation.

Peko smelter, Tennant Creek.

The first smelter was a failure, beaten by the twin problems of a depressed copper price and technical difficulties with top blowing matte in a rotary converter. Oil consumption was formidable and bismuth in copper matte could not always be held below the mandatory 0.04 per cent level. Operation of the rotary converter was cumbersome and inefficient. The smelter was closed down in February 1975, after 13 months of operation.

Warrego mine, Tennant Creek.

Some good fortune then came Warrego's way. One section of the orebody proved to be high in gold, with grades exceeding 20 grams per tonne, and this material was mined selectively, seeing the operation through the difficult times being experienced with copper. Some 40 tonnes of gold were recovered from the Warrego mine until its closure in August 1989.

The challenge of smelting and total control of bismuth on site remained. After four years of research into the problems, both in Australia and overseas, the company decided in early 1979 to recommission the copper smelter, in conjunction with fluid bed sulphation roasting to separate copper and bismuth. This would finally lead on to production of bismuth oxychloride and bismuth bullion.

After tests in the Mount Morgan smelter, a key decision was to replace the top blown rotary converters with conventional Pierce-Smith converters. The smelter was recommissioned and produced a copper matte with a satisfactory bismuth content. However, a depressed copper price once again put an impossible burden on this small smelter (20,000 tonnes per annum of blister was the objective), coupled with a small ore reserve. The smelter was closed down for the second time in October 1981.

Peko's operations continued, producing copper concentrate from the Gecko mine for treatment at Mount Isa and bismuth oxychloride from stockpiled material. After the smelter closed, a sixth mine, Argo, a small gold deposit at depth, was successfully mined for three years.

From 1954 to 1990, the combined operations of Peko's seven mines produced about 275,000 tonnes of copper, 2,560,000 ounces of gold, and 13,600 tonnes of bismuth. Employment peaked at about 1,100 in 1974, with another 200 people employed in other mining activities not associated with Peko in the field. Peko activities, along with its parent Peko-Wallsend Ltd, were merged into North Broken Hill Peko Ltd in the 1980s.

The Tennant Creek field was never a major feature on the map of Australia, but the many people from many nations who have worked in its sometimes extremely harsh and difficult conditions can be proud indeed of their achievements. The field has been a laboratory for operating small underground mines, for the practical application of rock mechanics, for the metallurgy of bismuth, for geophysical prospecting for hidden mineral deposits, and for efficient ore haulage by road.

I am indebted to Ray White, former General Manager at Tennant Creek and later Executive Director – Mining for the Peko Wallsend group, for the preparation of this paper, which was presented in Tennant Creek on 11 October 1990.

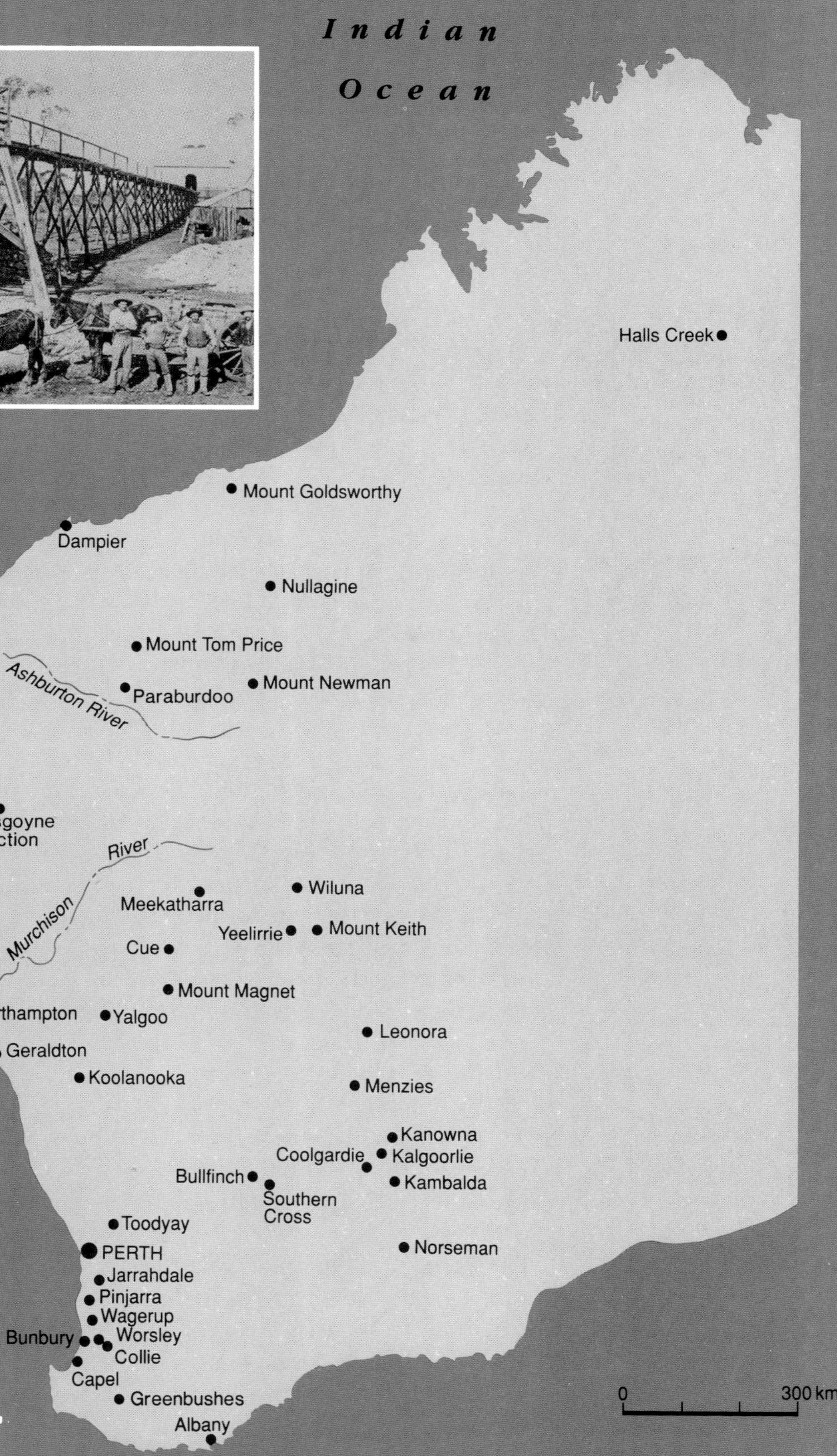

Western Australia

Safety in the Mineral Industry

The earliest mining in Australia was done by convict labour in the coal mines near Newcastle in 1797. Ross, in the *History of the Miners' Federation of Australia* records that the miners:

> slept naked in any part of the workings, the heat being so excessive that any clothing or covering only added to the misery of life. The convict miners remained underground the whole week and on Saturday afternoons were taken to the surface to wash themselves and their clothing in sea water.

It comes as no surprise that there are no safety records from this period.

The birth of Australian metalliferous mining in South Australia in the 1840s was different. The Cornishmen who came to work there brought with them mining technology, the tribute system, the social order of Cornwall, and safer working practices.

The earliest mining safety legislation in Australia was an Act passed in New South Wales in 1854, one of the objectives of which was 'to prevent accidents', but in this it had no marked success. Fatalities in coal mines occurred at the rate of around four per 1,000 man-years worked.

In Victoria, the transition from shallow alluvial gold diggings to deeper mines in the 1860s saw a worsening safety record. Between 1860 and 1870 an estimated 200 men lost their lives falling down shafts. Ladders were fixed vertically to the walls of shafts and the absence of staging accounted for most of these accidents.

The advent of wire rope and safety devices on cages in the late 1860s improved safety at some mines, but the long hours of winder drivers, careless shaft practices, and poor maintenance all led to the high accident rate continuing. Working garb consisted of felt hats, woollen singlets, jackets, bowyanged trousers and boots, each man carrying two candles for his shift. Some safety-conscious Cornishmen wore hard resinous felt hats and iron capped boots.

'A catastrophe in the mine. The rescue.' Engraving from a 19th century illustrated paper. Miners often had to climb ladders in the shafts, and falling down them was a common cause of death or injury.

A Victorian Royal Commissioner in 1863 declared that most deaths in the industry were caused by negligence or ignorance. A mine manager of the day commented, 'you have almost to knock [the miners] down with a stick to keep them from the [powder] magazine with a pipe stuck in their jaw'. But poor illumination, limited ventilation, bad ground, volatile explosives, and careless managers unquestionably contributed.

Low's patent rock drilling machine was first used in Victoria in 1867, but mechanical drilling was not common until the 1880s. This, and the wetter nature of the Victorian deep lead mines meant that silicosis, 'the miners' complaint', was not yet a serious problem in Australian mines.

The Victorian Regulation of Mines Act of 1874 was the most advanced for its time, preventing women and boys from working in mines and limiting the working day to eight hours. It put Victoria ahead of the other States in mine safety legislation. Important features were the appointment of inspectors and putting the onus on mine owners to prove that they were not negligent in mine accidents. A fatality rate of 3.3 per 1,000 man-years worked was recorded in Victoria in 1875.

Advertisement for Nobel's dynamite and detonators. Accidental explosions were a major mining hazard.

1880-1920

Despite legislation, by 1885 conditions in the coal mines of New South Wales and Queensland had improved little. A Colliery Deputy, Bill Renner, described the working conditions of the 1880s:

> There was not ventilation in the mines like there is today. The means of ventilation was by furnace which only kept the main road clean. I have known miners to be in the mine for a week at a time and their wives and children would bring their tucker and can of tea. They would be sent down the mine in cages. If there were no cages, most of the shafts had ladderways and the miners' wives would go up and down the ladders carrying their children and babies on their backs and their husbands' tucker.

The working environment was cramped, hot, and dusty. Coal mining also had the safety problems of flammable, toxic, and asphyxiating gases, the tendency of some coal to spontaneous combustion when exposed to air, and the presence of coal dust which in certain circumstances could be highly explosive. The use of explosives to break coal, candles for illumination, and furnaces for ventilation provided the potential sources of ignition to a veritable powder keg of combustibles.

The Bulli Colliery disaster of 1887, in which 81 were killed, was attributed to a methane explosion, set off by shot-firing which spread through the mine as a coal-dust explosion. Men were killed either by the explosion, or by the deadly 'afterdamp' (carbon monoxide) that followed. The worst Australian mining accident in terms of lives lost occurred at the Mount Kembla Colliery in 1902, when 96 were killed and 14 seriously injured.

In metalliferous mining during this period, the Creswick gold mine flooding in 1882 in Victoria, caused by the unexpected breaking into the flooded workings of a neighbouring mine, killed 29 men.

"Gunpowder, Cap'n! There ain't a man had more experience widdun."

This Oswald Pryor cartoon makes fun of accidents caused by dynamite, but they were not funny when they happened.

Six coal mine disasters in 15 years brought the perils of coal mining to public notice. The official inquiries that followed were remarkably similar in their conclusions and their condemnation of existing practices. Naked lights, poor shot-firing practice, smoking, accumulation of dust, inadequate ventilation, and inefficient rescue methods were among the problems highlighted. The New South Wales inquiries and a Royal Commission in Queensland in 1900, combined with rising union concern, led to a series of changes to the Mines Regulation Act.

Though statistics are scant and unreliable for this period, it appears that the stream of legislation and regulation had little impact on the safety record. Work practices changed little, although surface fans began to replace furnaces for ventilation, and 1904 marked the first use of coal cutters in Australia.

Mining lamp collection at the School of Mining Engineering, University of New South Wales. The combination of naked flames (such as candles) and explosive gases in mines was often lethal. Most modern mines use electric lamps powered by rechargeable batteries.

Rescue efforts in mines following accidents were at best ad hoc. Not until the 'Proto' apparatus was introduced in 1907 could mine rescues be conducted without risking the lives of the rescuers.

On the positive side, there was the remarkable rescue in 1907 of Modesto Varischetti from the Westralia and East Extension Mine at Bonnievale in Western Australia after nine days trapped in an air pocket up a rise when an overflowing creek flooded the mine. Three divers took food and candles to Varischetti daily for over a week, until the water level finally dropped enough for him to be carried from the mine.

The start of mining at Broken Hill in 1885 and at Coolgardie and Kalgoorlie in 1892 and 1893 resulted in an increase in mining accidents and fatalities, no doubt partly because of an influx of inexperienced workers.

Square set timbering with oregon timber was introduced at Broken Hill in 1888, and later at Mount Morgan in Queensland. The sets provided protection from local rock falls and the 'talking' of the timber gave warning of major rock falls or 'creeps'. Filling stopes with waste rock, and later by hydraulic fill, helped stabilise the workings.

Graphic as major accidents with loss of life in that period were, the real villain was occupational disease.

Dust from dry drilling, blasting, and dry underground filling was aggravated by the poor ventilation in working places. In an attempt to suppress the dust from drilling, an offsider would occasionally throw water on the hole.

A major contributor to the problem was the poor ventilation. The first recorded use of mechanical ventilation in Australia was the Roots Blower in Victoria in 1880, but under-capitalisation and a shortage of qualified engineers meant that available ventilation technology was not adopted uniformly.

Respiratory diseases were investigated by numerous Boards and Royal Commissions from about 1902. With the advent of X-rays, the cause of

Miners underground at Cobar about 1900. They used candles for light. Caps and hats kept the dust out of their hair, but did not protect them from injury from falling objects.

'Proto' men, wearing rescue apparatus, Broken Hill, about 1925.

Boiler explosion damage, Moonta Mines, South Australia, 1880s. Boilers provided steam for dewatering pumps and other equipment, but they generated high pressures and a blocked valve or pipe could produce a devastating explosion. Modern mines use electricity for most power needs.

Great Fingal mine, Western Australia, about 1910. Proper timbering added to mine safety.

Broken Hill's pulmonary diseases was found to be dust. A Technical Commission of Inquiry was set up in December 1919 to ascertain the facts 'using clinical and radiological means'. Headed by a pharmacologist, Professor H. G. Chapman, this Commission made, for the first time in Australia, systematic use of X-rays as a diagnostic measure in the investigation of an occupational disease. They concluded that pneumoconiosis caused by dust exposure in mines was in fact the main lung disease in Broken Hill mines.

Broken Hill had the added risk of lead poisoning and it was not until the late 1890s that open-cut mining of the carbonate orebodies, and improved ventilation underground, significantly reduced the incidence of 'leading'.

North Broken Hill mine, after flooding. Flooding was a constant fear in mines, either from overflowing streams on the surface, or from nearby workings.

James Hebbard, manager of the Sulphide Corporation at Broken Hill (and President of The Institute in 1913), had a concern for industrial hygiene which made his plant the most advanced in Australia. He minimised dust exposure by using exhaust fans, rotating jobs throughout the plant, and forcing men to wear respirators. Wet processing of ores and the use of flotation at Broken Hill from 1910 further reduced the dust hazard in concentrators. The first baghouse was built at Port Pirie in 1909 to trap the lead dust issuing from the smelter stack.

At the turn of the nineteenth century candles were still the main source of light in mines and poor light was no doubt a factor in many accidents. Candles gave way to carbide lamps, in coal mines to safety lamps. Electric lamps were introduced in the 1920s. Curiously, the Miners' Federation at first resisted them in coal mines 'because of their reduced illumination', ignoring the obvious safety benefits of replacing naked lights.

1920-1950

From 1920 to 1950, mining safety and occupational health standards improved markedly with technological, operational and management improvements. The impetus for those improvements was legislation and union activism. New technology, better methods, and improved processes made them possible.

The improved methods in all states resulted in the virtual disappearance of lead poisoning and lung disease from mines. Hollow drill steels for water flush, watering down practices, improved ventilation and fixed firing times all contributed to this. Workers' compensation legislation in all states and its extension to include designated occupational diseases gave companies a financial incentive to overcome occupational diseases.

Lead poisoning, though no longer present in the mines, was still a hazard in the smelters. Despite early efforts to improve working conditions at the BHAS Port Pirie smelter, there were 261 cases of lead poisoning in 1925. Mount Isa Mines Ltd also had a serious problem with lead poisoning in the late 1920s. It was lessened by flotation, but the solution lay in good housekeeping.

Flotation at the Zinc Corporation, Broken Hill. Wet processing greatly reduced dust problems in the industry.

The national average mining accident fatality rate fell from 1.5 per 1,000 man-years worked at the turn of the century to 1.2 per 1,000 man-years worked during this period.

Hard hats were not common until the 1930s. Despite the obvious safety benefits miners resisted improved personal protective equipment. Miners at Tennant Creek were still wearing felt hats in the 1960s.

In the period from 1920 to 1950 safety management and Mines Rescue organisations became formalised. The first comprehensive organised safety programme was at BHP's Newcastle steelworks in 1921. Formal

Dust collection underground at Broken Hill in 1920 was part of the Technical Commission's work into occupational diseases.

safety committees, accident investigation and reporting, and a clearly defined safety policy resulted in a 40 per cent fall in the accident rate. In 1925 George Dey became the first safety officer at Broken Hill Associated Smelters in Port Pirie, and in 1951 was awarded Honorary Membership of The Institute. The first safety department at Broken Hill was set up in 1936. Other mining centres followed.

Tipping molten slag at Port Pirie, early 1900s. Fumes from smelting operations at Port Pirie have long since been controlled.

Underground at Mount Isa, 1932. These miners would have been amongst the earliest to wear hard hats in Australia.

1950-Present

A great improvement in safety occurred in the post Second World War period. The accident fatality rate in the 1980s was a quarter of the prewar level, an improvement mostly brought about by technological change and better safety management. If the history of mining safety in Australia has a message, it is that safety legislation comes a poor second to good safety management.

The main technological change was the use of highly-productive earthmoving equipment that has made open-cut mining and mechanised underground mining possible on a large scale. Although this equipment brought new hazards, the net result is a safer industry.

In underground coal mining, mechanised coal cutting and loading in the 1950s brought greater productivity and improved working conditions. At first the Miners' Federation resisted mechanisation on safety grounds, fearing that faster rates of advance would increase the dust hazard and the danger of roof collapse. Mechanisation did involve some new hazards, but it also contributed to a marked reduction in accidents and fatalities in the 1970s. Engineered solutions were applied to accident prevention, overhead canopies and hydraulic chocks providing miners with unprecedented protection from roof falls.

Blasting as a cause of accidents and initiator of methane or coal dust explosions was all but removed from underground coal mining by coal cutting and loading technologies. Equally, the development of explosives in the 1960s and the gradual improvement in detonator technology from fuse to electric detonator to the non-electric detonator, reduced explosives-related accidents in metalliferous mines. In metalliferous mining, mechanisation such as raise boring and jumbo drilling minimised risks from freshly exposed rock, a major cause of accidents and fatalities. Introduction of the rockbolt during the 1950s was also a major advance.

Personal protective equipment has improved greatly, partly in response to the new hazards created by the mechanical equipment, but also as more-effective protection against dust and noise. The carbon monoxide self-rescuer is now widely used in both metalliferous and coal mines. After the 1979 Appin Colliery explosion in New South Wales, a deputy and nine men travelled almost a kilometre to safety using their self-rescuers.

It is a credit to the managers and technical people within the mining industry since the Second World War that new technologies have resulted in a net improvement in safety and occupational health. An important contributor has been the dedicated research done by Australian Coal Industry Research Limited and similar organisations. Of its ten research

Wyee State Coal Mine, New South Wales. Modern mining machinery has greatly reduced accidents underground, as well as improving productivity.

objectives set in 1986, nine relate either directly or indirectly to occupational health and safety.

Of particular note is the growth of rock mechanics as a tool for mine and ground support design. Quantitative analysis and prediction of rock behaviour became possible following the development of instruments for measuring stress. The standard instrument for stress measurement in underground mines is now the hollow inclusion stress meter developed by the CSIRO in Australia in the early 1970s.

To be effective, it is essential that safety programmes involve employee education to create the right attitude to safety. Active safety management is now a high priority for Australian mining companies, and the result has clearly been an improved safety performance.

Golden Crown Mine, near Cue, Western Australia. Modern drilling methods are safer and healthier than those once used.

Current Situation

While numerous accidents with large loss of life have occurred in the Australian mineral industry over its history of more than 150 years, such tragedies are not unique to the mineral industry.

In Australia, 89 people drowned in a flood in 1852 at Gundagai in New South Wales, 300 people lost their lives in a cyclone in 1899 at Bathurst Bay in far north Queensland, 66 lives were lost in Darwin in Cyclone Tracey in 1974, the Black Friday bushfires in Victoria in 1939 claimed 71 lives, and fires near Hobart in 1967 killed 62 people. Thousands have died in shipwrecks, aeroplane, train, bus and car crashes.

This is not an excuse for accidents in mining, but it is important to see the industry in perspective. Virtually all human activities, including the simple action of crossing the street, carry the risk of accidents and

Remote control front end loader underground at Mount Isa, 1989. Remote control enables operators to keep well away from possible hazards.

Rockbolting undergound at Mount Isa. Rockbolting prevents dangerous roof falls.

loss of life. The important question is: what is being done to minimise the risks?

In Western Australia the Mines Regulation Act and Regulations, which have evolved over 100 or more years, provide a sound basis for the operation of the industry based on the co-operative participation of mine management and the workers, and a regulatory framework involving a skilled inspectorate.

During the 1980s the gold mining industry in Western Australia grew from three major mines employing about 1,640 people and producing 350,000 ounces of gold a year, to 66 significant mines employing 9,100 people and producing 4,350,000 ounces a year. Such growth placed an enormous strain on the availability of skilled people in all sectors of the industry. There was a desperate shortage of experienced management and many inexperienced people joined the workforce.

The Mines Department had difficulty in obtaining suitable inspectoral staff, and this placed an intolerable load on an already stretched Department when the need for its expertise was perhaps at its greatest. Despite these difficulties the lost-time injury incidence rate in the gold mining industry in the period 1981 to 1989 was reduced from 210 to 105 per 1,000 employees per year.

On a number of measures recent Western Australian Government statistics show that the mining industry safety record is better than that achieved by several other industries including transport and storage, public administration, and manufacturing.

Indicators of safety performance have had to be revised over the years to express the greatly improved results. Whereas the measure was once fatalities per 1,000 employees per year, later measures report lost time injuries per million hours worked.

Some near accident-free installations have adopted an even finer measure, the frequency of medical attention given per million man hours. Even using this measure, there are now some remarkably low rates.

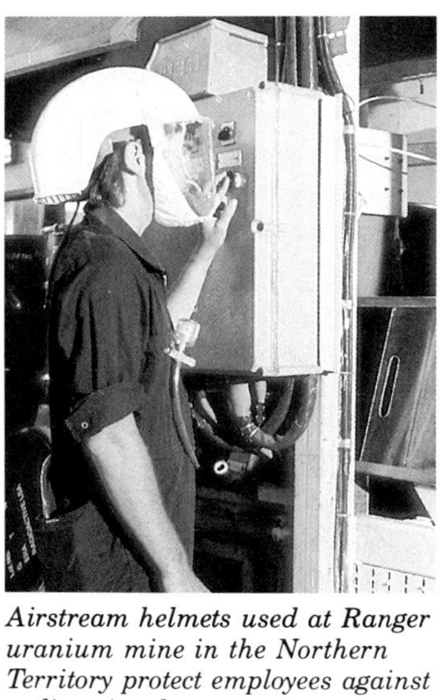

Airstream helmets used at Ranger uranium mine in the Northern Territory protect employees against radioactive dust.

Conclusion

The trend in safety performance is clearly for the better. While, regrettably, there are occasional setbacks, the mining industry can be proud of what it has achieved. The challenge remains for further improvement. Those in the industry know that a great, and sustained, effort is being made. The ultimate aim is an accident-free working environment. Nothing less than this can be our objective.

I am indebted to Ross Muller and David Rose of WMC Nickel Divison and Gilbert Ralph, MBE, for their invaluable assistance in the preparation of this paper, which was presented in Perth on 4 October 1990.

Gold Brings Prosperity to Western Australia

Influenced by the prosperity resulting from gold finds in eastern Australia, the Western Australian Colonial Government in 1862 offered a reward of £5,000 for discovery of a workable goldfield in Western Australia. This attracted the attention of Edward Hargraves, who had discovered gold near Bathurst, New South Wales, in 1851. For a fee of £500 plus expenses he came to Western Australia and, on arrival at Albany in October 1862 declared: 'I do not think I will pass within five miles of a goldfield without discovering it'. His 1863 report on his journey to the Murchison and back concluded that he, 'did not think that a goldfield existed on or anywhere near his track'. Despite this, H. Y. L. Brown was appointed Government Geologist in 1870.

The discovery of gold in 1886 at Halls Creek in the Kimberleys brought about the first Western Australian rush. Further discoveries were made as the prospectors moved south through the Pilbara and Murchison.

Henry Maxwell Lefroy. In 1863 he led the first exploring expedition through the area that later became the Eastern Goldfields of Western Australia.

The Yilgarn

In 1887 a prospecting party consisting of Richard Greaves, Edward Payne and Harry Anstey found gold at Eenuin in the Yilgarn. Greaves and Payne continued to prospect in the area under great difficulty, and Greaves recorded that:

> In our half blind state we had left our small bag of flour at a camp, and with no water and no damper, and no prospect of getting any more water in this part of the country, we were indeed in a bad plight. The pain of our inflamed eyes was simply terrible, and very depressing to our spirits, as there was a possibility of our never getting back to the Eenuin camp. We had already been 63 hours without a drink, but our lives were saved by our native guide. . . . He gave us both food and water.

Back in Perth, Greaves met two prospectors, Bernard Colreavy and H. Huggins, and told them of the location. Colreavy and Huggins went there, found the main lode and called the place Golden Valley. Southern Cross, just south of Golden Valley, grew rapidly after Toomey and Riseley found reefs there in January 1888, and Parker found gold in Parker's Range. The rush to the Eastern Goldfields had begun, with Southern Cross to be the centre of activities for the next three years.

Camels proved to be a useful means of transport in the dry, sandy Eastern Goldfields country.

Coolgardie

When in 1892 the mine owners at Southern Cross announced a wage reduction, many miners left in search of more rewarding prospects. Among them were Arthur Bayley and William Ford, who ventured along Hunt's track to Gnarlbine Rocks, then north-east towards Mount Burges,

Dry blowing on the Eastern Goldfields of Western Australia. Dry blowing was a way of separating gold from earth when there was no water.

to discover over 500 ounces of gold in a matter of hours at Fly Flat, Coolgardie. Bayley returned to Southern Cross to register a reward claim.

When the news broke, it started what Keith Quartermaine and Ted McGowan described, in their excellent summary of mining developments in Western Australia in the sesquicentennial volume *Mining in Western Australia*, as, 'the most impulsive and intensive rush to date. It left Southern Cross a deserted shanty town, the strike bound mines forgotten, and a week or so later the rush was reinforced from the coast and from other fields'. For a year or more eager prospectors swarmed to the area from all over Australia, and from overseas. For a few years Coolgardie prospered despite the fact that a number of the richest discoveries did not live up to expectations.

Statue of Paddy Hannan, Kalgoorlie. Many a school boy has drunk from his water bag since it was erected in Hannan Street in 1929.

Kalgoorlie

Good rains in mid-1893 prompted many to leave the overcrowded Coolgardie field, among whom were Patrick Hannan, Tom Flanagan and Dan Shea. They followed a rush to a reported find at Mount Youle, 80 kilometres north-east of Coolgardie and en route stopped at the low range of hills where they found alluvial gold on or about 15 June 1893 at Mount Charlotte, near Kalgoorlie. At first the find created only modest interest. Not until after the discovery of the Golden Mile lode system by F. W. P. Cammilleri at Block 45 and the rich Boulder leases by William Brookman and Sam Pearce later in 1893 was there much attention and investment directed to Kalgoorlie.

Neither Hannan nor his mates made a fortune. They sold the claim and moved on within a few months, considering the area too crowded. In 1908 the Western Australian Government granted Hannan and Shea annuities of £100 each for life. Hannan went to Melbourne where he died in November 1925 a poor man.

The story of the development of the Golden Mile at Kalgoorlie-Boulder is well known. Many prospectors, for various reasons, fanned out from existing finds in search of the elusive 'El Dorado'. Their efforts led to Kanowna (1893), Menzies (1894), Norseman (1894), Leonora (1896), and Kambalda (1897) being discovered and developed.

In October 1893 a party of prospectors came into Coolgardie with 50 ounces of gold from 110 kilometres north. They said there was no water on the track. 'But there's gold!' was the excited reply and away went dozens of determined prospectors. They met others returning from the find who said: 'It's suicide to go on. There isn't any gold and there isn't any water!' Several ill-prepared parties kept going. Malcolm Uren records that Charles Renou, the Government water carrier, saved many from certain death. Uren wrote: 'The trail was strewn with abandoned swags. Renou's men buried ten of them, those who died on the backward trail. Others, maddened by thirst, wandered off the trail, stripped their clothes, as thirst-crazed men do, and died in the desert'. The place was named Siberia.

On the Golden Mile, about 1895.

On the Golden Mile, 1902.

Consolidation

The impact of these Eastern Goldfields discoveries on Western Australia was enormous. The population, about 29,000 in 1890, grew to 110,000 in a decade, Many people were attracted from the eastern colonies, where there was a general depression. Much of the capital for the Kalgoorlie field came first from Adelaide, later from London. The complexity of the telluride ore and increasing depth of the mines required capital.

Bayley's No. 1 South mine, on the site of Bayley's find, Coolgardie.

Power House at Lake View and Star Mine, Kalgoorlie, early 1930s.

Western Australia's production reached a peak in 1903 at 2,064,803 ounces and for the next 80 years Kalgoorlie's annual production exceeded that of any other centre in Australia. However, it was not until 1984 that the cumulative gold production in Western Australia exceeded that of Victoria, despite Victoria's production having dwindled from a peak of 2,875,647 ounces in 1856 to 631,986 ounces the year Kalgoorlie was discovered and having largely ceased by 1920. By 1990, Kalgoorlie's uninterrupted gold production had exceeded 42,000,000 ounces, making it by far the biggest producing field in Australia.

Norseman

Norseman is another field with an almost uninterrupted record of production from its discovery in 1894 with nearly 4,500,000 ounces, second only in Western Australia to Kalgoorlie. Gold was first found at Dundas in 1892 by a man named Moir, who was searching for pastoral land. In August 1894, the Norseman Claim, 22 kilometres north of Dundas, was registered by Lawrie Sinclair, on behalf of a five-man prospecting party. 'Norseman' was his horse. The Norseman reef was soon the centre of feverish activity, particularly after the rich Mararoa and Princess Royal reefs were discovered. The field reached its zenith in 1910, after which there was a steady decline, and by 1920 there were fewer than 300 men on the field, mostly tributers.

The field was reopened in the mid 1930s by Norseman Gold Mines N.L. and Central Norseman Gold Corporation N.L., the latter company being a part of the Western Mining Corporation Group. It successfully applied new methods of exploration and development. Of mines producing gold in 1990, Central Norseman has been operating longer than any other in Australia.

Frank Moss, President of The Australasian Institute of Mining and Metallurgy in 1907.

Services to the Goldfields

Success for the Eastern Goldfields of Western Australia depended on reliable services. The Western Australian Government at its own expense cleared the tracks, made roads, established water supplies, extended telegraph and postal services, built railways, and erected substantial public buildings, many of which remain as distinctive symbols of the State's mining heritage.

The Goldfields Water Scheme was the most ambitious and costly undertaking. Whilst there is some doubt as to who conceived the idea, it was designed and constructed under the guidance of C. Y. O'Connor, the State Engineer-in-Chief. When opened in 1903, after six years of planning and construction, it was the longest pumped water scheme in the world and was regarded as an engineering marvel. Regrettably O'Connor, who was subject to a great deal of public criticism about its practicability and cost, was so disturbed by the attacks that he took his own life some months before the scheme was completed.

Dredging for alluvial gold began at Coolgardie soon after the Goldfields Water Scheme was commissioned, but it proved unprofitable and the dredge was relocated to the Greenbushes tin field, in the south-west of Western Australia.

The fabulous Golden Mile, Kalgoorlie, as recorded by Western Mining Corporation's air survey unit in 1933 showing the Associated mine at top, Kalgoorlie Ore Treatment Co in the lower centre and the Great Boulder mine at lower right.

Heroes and Heroines

There have been many examples of heroism on the Goldfields, not the least of which occurred at the Westralia and East Extension Mine at Bonnievale north of Coolgardie in March 1907 when a flash flood filled the lower levels of the mine, trapping Modesto Varischetti in a rise off the 10 Level. When rescuers tapped on the level above, Varischetti responded. How long could he survive, partly immersed in the cold water, without food and air?

Family life on the Eastern Goldfields in the early 1900s often meant life in a bark hut.

Divers and diving equipment were rushed at record speed to Coolgardie by train. When a visiting diver was injured on his first attempt to rescue Varischetti, Frank Hughes, a miner from the South Kalgurli Ltd mine, who had some experience at diving, ventured down. He made contact with Varischetti, and supplied him with food, an improvised battery light and fresh air from a hose. Such visits continued daily until the ninth day, when the water level had been lowered enough by bailing that Hughes was able to bring Varischetti out of the mine to safety. Hughes received an elaborate illuminated address which is now in the Museum of the Goldfields.

Many women went with their husbands to the newly-discovered goldfields and shared the excitement of adventure and discovery, as well as the many hardships. Twenty-two year old Wilhelmina Sloss arrived in Hannans with her husband and infant daughter within a few weeks of its discovery to become the first woman in the area. She later helped nurse miners during a typhoid epidemic.

> At first (the) women lived in a tent or hessian hut, a stamped earthen floor inside and a camp oven outside, packing cases for chairs and hessian slung between poles for beds. When they moved to houses of corrugated iron and white washed walls they thought they were aristocrats, but when typhoid raged and neighbours were ill they were Samaritans.

Jack Fimister, after whom the Fimiston mining area near Boulder was named. He was at one time Mayor of Kalgoorlie.

Peaks and Troughs

Western Australia's gold production declined steadily from 2,069,803 ounces in 1903 to 377,176 ounces in 1929. The State had fewer new discoveries, the ore grades of established mines declined, small fields were depleted, and costs increased, while the gold price remained static.

During and after the First World War many companies which otherwise would have had to close survived by letting parts of their leases to tributers, parties of miners who paid a tribute on the amount of gold they won. By 1925, the Sons of Gwalia mine at Leonora was the only mine outside the Golden Mile producing over 12,000 ounces per annum. The Sons of Gwalia mine became outstanding in Western Australian mining history in that it operated continuously from discovery in 1896 to 1963. At the turn of the century it was managed by Herbert Hoover, who later became President of the United States of America.

During the 1920s the once-proud communities of Kalgoorlie and Boulder, with their imposing Town Halls, in which banquets had been held to entertain Prince Edward, Major-General Sir William Birdwood, Prime Ministers, and a host of stars including Dame Nellie Melba, GBE, and Clara Butt, were falling into disrepair. The Western Australian School of Mines, which had commenced at Coolgardie in 1902 and transfered to Kalgoorlie in 1903, struggled to survive. All but one of the dozen or so breweries closed and the number of hotels was reduced from nearly 100 to less than half that number. Coolgardie, once a vibrant place which hosted the Western Australian International Mining and

Kalgoorlie Miner

ESTABLISHED 1895

TELEPHONE: KALGOORLIE 21 2122, ESPERANCE 71 4343

The Kalgoorlie Miner, established in 1895, is one of several Australian newspapers that elected to name themselves after the majority of their readers.

Industrial Exhibition and the renowned Westral Wheel Race, was all but a ghost town. By the end of the decade, the major department stores of Hicks, Brennan Bros and Montgomery Bros. were in decline.

Despite the depression, the 'Dirty Acre' at Fimiston, which had a brewery and six hotels (including the Boulder Block, which was still trading in 1990), and a variety of other places offering 'entertainment', remained active. Charlie Cutbush, one time Mayor of Kalgoorlie, claimed to have acquired the surface rights of the whole block from W. G. Brookman for a case of champagne. J. W. Fimister, who gave his name to Fimiston and was later Mayor of Kalgoorlie, once owned most of the area.

Town Hall, Boulder, about 1930.

Government Support

The plight of gold mining became a matter of concern and with the onset of the Depression from 1929 the industry, led by the flamboyant Claude de Bernales, launched a campaign seeking a £1 per ounce gold subsidy for increased production over the 1928-29 level.

The Gold Bonus Campaign was successful and de Bernales was appointed Chairman of the Gold Industry Advisory Council set up by the Government to administer the bonus. Before the scheme could be implemented, however, other events overtook it. Britain and Australia went off the Gold Standard and the Australian pound was devalued by 25 per cent. The gold price rose from £4.5.0 per ounce in 1929 to £8.10.6 per ounce in 1934.

The Rich and the Famous

Claude Albo de Bernales was probably the most colorful and controversial character associated with the Western Australian Goldfields. He arrived at Coolgardie in 1897, aged 21, and soon had a reputation as a dapper salesman of second-hand machinery. He charmed, and later married, the widow of the proprietor of the Kalgoorlie Foundry and went on to build an empire of inter-related gold-mining and machinery-supply companies.

Claude Albo de Bernales, Western Australian gold-mining promoter.

Later he took part in forming Wiluna Gold Corporation and was Chairman of several other London-based mining companies with operations in Western Australia. There was little long-term value in any of de Bernales' mines except Wiluna, and the network of companies he created were marvellously interlocked to his personal advantage. In 1939 the

Central Norseman Gold Corporation Ltd Phoenix mine, Norseman, 1940.

London Stock Exchange suspended all trading in his shares and one by one his companies were ordered into liquidation.

During the Second World War he became Chairman of the Great Boulder Company, which he tried to misuse for personal gain. This led to an angry response from shareholders and investigations by the British Treasury and Scotland Yard. A great deal of evidence was assembled against him and when the matter finally went to Court in 1947, de Bernales was said to be 'too ill' to appear. He died in 1963, aged 84, without the case being heard.

During the revival of prospecting in the 1930s depression a 16-year-old boy, Jimmy Larcombe, working a claim with his father at Larkinville, uncovered a lump of gold which weighed 1,134 ounces. The 'Golden Eagle' remains the largest nugget ever found in Western Australia and its finding attracted thousands of prospectors and fortune hunters to the region.

Frederick G. Brinsden, a Victorian metallurgist who went to Kalgoorlie in 1902 to work with Kalgurli Gold Mines Ltd, and spent all his working life in Kalgoorlie, was President of The Institute in 1933. He was awarded the Institute Medal in 1947 for 'his contribution to the metallurgical treatment of sulpho-telluride gold ores'. The problem of efficiently treating such ores had attracted the attention of prominent gold metallurgists from around the world, and this resulted in two processes, the wet bromo-cyanide process and the dry crushing-and-roasting process. Brinsden was involved in experimentation into these processes and improvements to them. He was General Manager of South Kalgurli Consolidated Ltd for the greater part of his time in Kalgoorlie.

The fabulous Golden Eagle nugget, found in 1931 by Jimmy Larcombe. This sensational find was the stimulus that brought farmer George Cowcill to the Kambalda area to try his luck. Much later, Cowcill found the nickel that led to the modern development of the Kambalda nickel field.

Joe Thorn, an American-born mining engineer, went to Kalgoorlie in 1929 to take over management of Lake View and Star Ltd at a crucial period in its history. He was awarded the Institute Medal in 1945 for his contribution to gold mining in Western Australia. He was responsible, amongst other things, for the introduction of flotation, roasting of the flotation concentrate and the introduction of diesel-electric power.

R. J. (Dolph) Agnew graduated from the W. A. School of Mines in Mining Engineering in 1928 and after a period working in England, Yugoslavia and Italy returned to Western Australia in 1932 and became General Manager of Lake View and Star after Joe Thorn. As President of The Institute in 1955 he delivered his Presidential Address in the

Kalgoorlie Town Hall on 16 August 1955 on 'The Development of the Western Australian Goldmining Industry':– an excellent resume of the major events in the industry up to that time. He highlighted the significance of the introduction of light drilling equipment, tungsten carbide drill tips and sharpening methods, locomotive haulage underground and flotation of sulphide ores.

Western Australia has provided other Presidents of The Institute, including Frank Moss in 1907, Richard Hamilton in 1909, Fred Brinsden in 1933, Frank Espie (Snr) in 1948, Dolph Agnew in 1955, Edgar Elvey in 1964, Sir Laurence Brodie-Hall,CMG, in 1973 and Dr Don Zimmerman in 1987, as well as a number of Institute Medal winners.

Revival

During the 1930s, British capital again flowed in and the revival of the industry eased economic hardship and reduced unemployment. A flood of workers came to the goldfields to take part in the expansion, building and re-equipping of gold mines. By 1939 Western Australian production had climbed back to 1,214,238 ounces and the work force had almost doubled.

It was during this period that a great deal of rebuilding of the Lake View and Star, Great Boulder and North Kalgurli plants took place and new ventures such as Gold Mines of Kalgoorlie Ltd, Central Norseman Gold Corporation N.L., Triton Gold Mines N.L. and Wiluna Gold Corporation Ltd emerged.

Almost as soon as the gold mining industry was enjoying the fruits of revival, it was again crippled by war. Many of the mines were forced to close as a greater part of the able-bodied men went off to the Second World War, some in a Goldfields Regiment. Recovery after the War was handicapped by the shortage of manpower and materials. Some out-lying

Machinery merchants continued to prosper on the Eastern Goldfields.

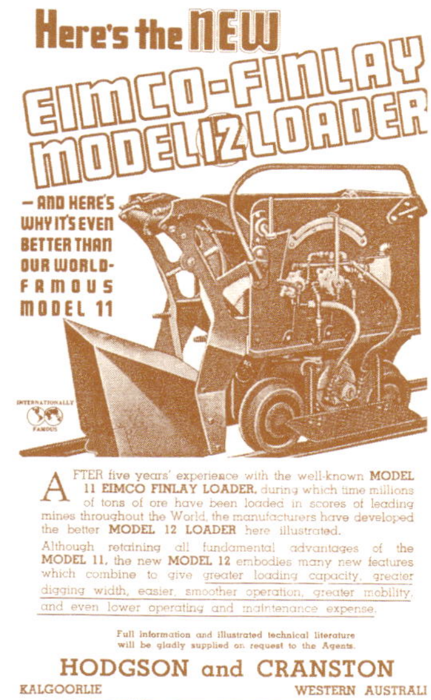

A prospector and his dolly pot. The Goldfields attracted many prospectors during the Depression.

mines never re-opened. Others emerged, including a number in the vicinity of Coolgardie and a major development at Bullfinch-Southern Cross-Nevoria in 1950 by Great Western Consolidated Ltd.

Silver Lake shaft, Kambalda, the first nickel mine in Australia.

Postwar redevelopment of the gold mines in Western Australia saw further rationalisation, when some of the smaller companies, such as Kalgoorlie-Enterprise Gold Mines Ltd, Paringa Mining and Exploration Co. Ltd, Boulder-Perseverance Ltd and South Kalgurli Consolidated Ltd were absorbed. Only four major producers remained: Lake View and Star, Great Boulder, North Kalgurli and Gold Mines of Kalgoorlie.

The gold price rose to £15.10.0 per ounce in 1949 when Britain depreciated the pound sterling. But the benefit of this was quickly eroded in the early 1950s by cost increases because of inflation and declining grades, and by 1954 a Federal Government subsidy was again needed to maintain operations and sustain employment, which had fallen from 15,000 in 1939 to 7,000 in 1949.

In the early 1960s the only successful new development was the reopening of the Mount Charlotte mine by Gold Mines of Kalgoorlie, using diesel powered equipment underground in large open stopes in Western Australia for the first time. By the latter half of the 1960s conditions in the Goldfields were depressed, and most companies were losing money. The two-tier price system for gold was of little help. Kalgoorlie, Norseman and Mount Magnet, the major communities reliant on gold mining, were severely affected.

Nickel Boom

In 1965 two part-time gold prospectors, George Cowcill and John Morgan, who had been working in the Kambalda area, brought some mineral specimens to Western Mining Corporation. They had been analysed at the Western Australian School of Mines some 10 years earlier and identified as containing minor nickel. WMC began diamond drilling in the area on 28 January 1966, and a drill operated by Jack Lunnon soon intersected a massive nickel sulphide ore body.

The discovery, announced in April 1966, began what became known as the 'nickel boom', a period of frenzied activity on the Goldfields, reminiscent of the gold rushes of the 1890s. More than 200 companies, syndicates and partnerships explored for nickel. As with the gold rushes, only a few survived.

Repairing a rod mill at Kambalda, 1986. Ore is crushed into a fine pulp in rod and ball mills to enable its metal content to be extracted.

One difference between the nickel boom and the gold rushes was in the government contribution. In the 1880s and 1890s the infant Colonial Government developed the essential services to encourage the newly-emerging industry, but in the 1960s the companies had to finance most of the infrastructure and make major contributions to essential services, even before they had any income from their operations.

The nickel developments included Australia's first nickel smelter at Kalgoorlie in 1972, which introduced pyrometallurgy to the Eastern Goldfields. Later the Uranium Research Plant to test Yeelirrie ore added chemical engineering. Among other features of this era were rapid advances in exploration technology, including a variety of geophysical

Miners in an underground crib room, Kambalda, about 1969.

and geochemical methods and much-improved analytical techniques, such as the atomic absorption spectrophotometer, which greatly enhanced the exploration processes through faster and more accurate analysis.

Consolidation for Survival

In 1973 Gold Mines of Kalgoorlie Ltd merged with Lake View and Star Ltd to form Kalgoorlie Lake View Pty Ltd, which in turn in 1976 formed Kalgoorlie Mining Associates, a joint venture with Homestake Gold Ltd, in an effort to survive. Elsewhere in Western Australia, Central Norseman Gold Corporation N.L. was the only other significant producer, Hill 50 Gold Mine N.L. having ceased production in 1976. The Telfer Mine, in the Pilbara, was the only new gold mine in Western Australia for more than a decade.

An increase in the gold price late in 1976 saved the Western Australian gold mining industry. Kalgoorlie Mining Associates reopened the Fimiston mines, North Kalgurli resumed gold production and later other producers entered the field, reopening old mines and developing new open-cut operations. Several changes of ownership and control brought most of the Kalgoorlie field under the single management of Kalgoorlie Consolidated Gold Mines Pty Ltd, with the obvious benefits of scale and reduction of overheads.

Converters at Western Mining Corporation's Kalgoorlie nickel smelter.

Sir Laurence Brodie-Hall, CMG, a graduate of the WA School of Mines in Kalgoorlie and former Director of Western Mining Corporation, devoted a great deal of his time to the expansion and development of the School after his retirement in 1975.

Revival of gold mining in the 1980s occurred not only because of the improved price of gold but also because of technological advances in exploration, mining and mineral processing.

Future Outlook

The history of 100 years of mining on the Eastern Goldfields spans half the history of European settlement of Australia. The contribution to the development of Western Australia and Australia has been irreplaceable.

While ups and downs in the level of activity and prosperity are characteristic of the industry, the important feature of the region is that it is a highly-mineralised part of Australia. More exploration is likely to result in more mineral finds. I have no doubt that mining will remain an important activity in the Eastern Goldfields for decades and perhaps centuries to come.

I am indebted to Gilbert Ralph, MBE, for his invaluable assistance in the preparation of this paper, presented in Kalgoorlie on 1 June 1990.

The Mount Charlotte mine in Kalgoorlie was the first underground mine in Western Australia to be equipped with diesel powered mobile units.

Murchison Gold

Western Australia's first economic mining venture resulted from the discovery, by A. C. Gregory, a Government Surveyor, of lead and copper in the bed of the Murchison River in November 1848. Governor Fitzgerald visited the site to settle a dispute regarding mineral ownership and the *Cyclopaedia of Western Australia* of 1913 records that, 'his trip resulted in bloodshed, owing to an attack by the natives – The Governor being himself wounded by a spear thrown at close quarters'. By April 1852 export of pig-lead had commenced and by June 1854, under the management of Captain Evans, a new smelter was operating at a daily rate of 1.5 tons of pig-lead. A massive stone chimney remains as a monument to this activity.

Dry blowing is one way of separating gold from dirt when there is no water for panning or puddling. Just as grain is separated from chaff by throwing the mixture into the air, the wind blows away the lighter dirt while the heavier gold falls to the ground.

Lead and copper were also found by a shepherd, Thomas Mason, at Northampton, 52 kilometres north of Geraldton, and this became the first significant mining centre in Western Australia. These mines were developed using English capital and worked to a depth of up to 100 metres. The *Cyclopaedia* reports 'the smallness of the veins, enclosed in intensely hard rock, made following them to greater depth expensive – for these were the days of black powder and hammer and drill work for blasting'.

The mine was the reason for the first railway line in Western Australia, between Northampton and Geraldton. Operations peaked in 1877, when the export value of lead reached 13 per cent of the colony's export earnings. After this the activity declined but the town and remnants of the mine remain as evidence of this early mineral history.

Robert Austin, the leader of an expedition in 1854 in search for grazing land, stood on top of Mount Magnet (which he had named because of its effect on his compass) and reported that 'the geological formation of the country east of the Murchison River has every appearance of being one of the finest goldfields in the world'.

The discovery of gold at Halls Creek, in the Kimberleys, in 1886, brought about Western Australia's first gold rush. From here, prospectors moved south and the first rush to the Pilbara began on Christmas Day 1887 when young Jimmy Withnell picked up a stone to throw at a crow, and found it to be flecked with gold.

Nathaniel Cooke found gold at Nullagine in 1888. It is a very harsh and arid region, Nullagine and Marble Bar being amongst the hottest centres in Australia, and it was not surprising that many of the prospectors moved south through the Ashburton and Gasgoyne to the Murchison.

Early Days of the Murchison Field

The Murchison Field was proclaimed in 1891, the year after John F. Connelly was rewarded for the discovery of gold in the region. Connelly, a surveyor and mining engineer, found a rich gold patch on Annean

Dry blowing for gold, Day Dawn.

Station, near Nannine, but was beaten in registering his claim by Gilles McPherson and J. Peterkin. They had read of Connelly's find in a newspaper, journeyed out to Annean Station, where Richard Robson, a jackeroo, gave them a map which Connelly had accidentally dropped at a gate.

Malcolm Uren describes how an errand of mercy led to the discovery of gold near the town of Mount Magnet in 1892. When a man on a drunken binge near Yowergabie Station lost his swag a station hand named Steadman took an Aboriginal tracker and retraced the drunk's steps to Boogardie, where Steadman picked up a two-ounce nugget. Whether or not the swag was recovered is not recorded. Others claim that the first lease granted at Mount Magnet was in August 1891. Considering that the hill called Mount Magnet is 40 kilometres from the town of Mount Magnet, such confusion is not surprising.

Soon after this Tom Cue found gold near the town which bears his name, and a succession of discoveries brought prospectors from many areas to the Murchison.

Meekatharra attracted attention late in 1895 when three prospectors, Meehan, Porter and Soych, pegged a claim near the present town. Most of the area's early gold came from mines and alluvial workings on Paddy's Flat. The largest producers were the Lakeview leases and the Ingliston

Extended Mine. Production from the Meekatharra area to 1987 was 1,290,000 ounces.

The Great Fingall mine at Day Dawn was one of the most successful on the Murchison Field, producing well over 1,000,000 ounces of gold with plant and equipment which was huge and amongst the most-efficient of its period. After producing 1,182,802 ounces from 1,862,593 tons and paying good dividends, it ceased production at the end of the First World War and never regained its former prominence.

P. R. Heydon in his book *Gold on the Murchison* describes how towns were built on the Goldfields:

> It is amazing how quickly the towns of Cue and Day Dawn were formed. All freight had to be laboriously hauled through the bush, along unmade tracks, from Mullewa to Cue – a

Headframe under construction, West Fingal mine.

Cue township, late 1920s. Tom Cue found gold in the area in the early 1890s.

distance of 200 miles. The only means of transport [were] camel, bullock, donkey, or horse teams. Some of the lucky few fossickers used pack horses, and luckier ones used buggies. But soon professional teamsters moved in and, in 1894, there were 400 teams carting from Mullewa to Murchison centres. They were mainly horse teams, but some long strings of donkeys were used. Faiz and Tagh Mahomet brought the first camels to the area, and although there was opposition to these ugly animals they soon reduced the cartage rate to thirty pounds per ton from Geraldton to Nannine.

The Afghan driver dressed in a dirty pair of baggy trousers which seemed to have been formed from a continuous roll of material, and a loose-fitting shirt that had not seen a washing tub for some considerable time. He had a swarthy, dry, sunburnt face, and on his head was a large turban, looking like a scrap bag, through sundry holes in which ends and ribbons protruded and fluttered free.

Big Bell head frame, winder house and power house.

The strangely named Youanmi mine in the East Murchison was discovered in 1896 by Thomas Payne, an experienced prospector who gave his name to Paynes Find and Paynesville. The origin of the name Youanmi is lost in history, some claiming it to be a corruption of an Aboriginal word Youan, which is not uncommon in the area, while others think it more likely that it is simply taken from you-and-me, as it is pronounced. Youanmi, which became a significant producer between 1911-22 and again 1936-42, was managed by the English firm Bewick,

Moreing and Co. during the former period. They had a big influence on mining on the Western Australian Goldfields during the first two decades of the century. To 1987 the Youanmi yielded 265,700 ounces.

Another of the great mines of the Murchison was discovered by chance. In 1904 an Aboriginal woman picked up a specimen near Cue and showed it to a prospector named Paton, who took from the site a parcel of ore which went two ounces to the ton. Later he lost the reef and abandoned the lease. It was taken up by James Chesson and William Hayden, and working it as an open-cut, they soon won 10,000 ounces. They called it Big Bell. Later they too abandoned it and the lease remained idle for many years until Herman Mandelstamm took it up. In 1937 the mine was reopened on a grand scale by an American company using the flotation process to treat the low grade ore. It closed in 1954 but was reopened in the 1980's and to date has produced over 741,000 ounces.

Train tipping into an ore chute, Big Bell, during the mine's 1937-54 operation.

Revival during the Depression

Big Bell was one of several new developments in the Murchison following the increase in the gold price when Britain abandoned the gold standard in 1931 and the Australian pound was devalued, causing the price to double.

During the Depression, investment in gold seemed very attractive for those with funds. One story goes that during a visit to one of the Murchison's more successful mines a shareholder asked which way the richest shoot was heading, no doubt with a view to further investment in an adjoining property. 'Well' the Manager replied, 'judging by the way the ground is pegged and the names allotted, I would say the reef runs in a circular direction'.

Claude Albo de Bernales had a notable impact on the Murchison area about this time. Of all the mines to which he attached British capital, Wiluna was the only one with any long-term value.

A contemporary account of the Wiluna Gold Mines Ltd said that, 'the management has been a model of well-applied energy and efficiency, and a policy of intensive development has been in operation. Large reserves of payable ore have been opened up and important new discoveries have been made. The property which comprises 1,168 acres [473 hectares], gives rise to the question whether the Company owns a mine or a goldfield'. The property was equipped with the most modern plant and equipment at a cost of £1,500,000. Production began in April 1931 at the rate of about 40,000 tons per month, with 300 men employed per shift. The Government extended the railway from Meekatharra to service the new development. With a production of over 1,770,000 ounces, Wiluna was the biggest producer on the Murchison.

Undergound surveyor at Big Bell.

At Reedy's, near Cue, Triton Gold Mines N.L., a member of the Western Mining Corporation Group, reopened the mine and established a new 100,000 tons per annum treatment plant which began operating in 1936. Typical of the gold mining industry at the time, it used a headframe and winders which had previously been installed at Bendigo, Victoria, and which were later transferred to the Copperhead Shaft at Bullfinch

Battery at the Champion mine, Nannine. While generally trouble free such plants were inefficient and very noisy.

and later again to the Durkin Shaft at Kambalda. Triton produced more than 300,000 ounces.

The main lode at Mount Magnet was also reopened during the 1930s when Hill 50 Gold Mine N.L. was floated in 1934. Two mines at Mount Magnet were named Hill 50 and Hill 60 to commemorate military objectives near Ypres in France during the First World War, because the 50 and 60 metre contour lines identified their crests.

The reopening of these and other mines in the adjacent Peak Hill, Yalgoo, Ashburton and East Murchison Goldfields helped to turn Western Australia's annual gold production around from 377,176 ounces in 1929 to 1,214,238 ounces in 1939.

The Second World War and its Aftermath

The Second World War had an even more serious impact on mining than the First. With gold not considered a strategic metal, many mines were forced to close for want of manpower and because the Government requisitioned some of the generating plant and machinery. The direct workforce in gold mining in Western Australia fell from about 15,000 in 1939 to less than 5,000. After the war some of the Murchison mines reopened but, apart from Hill 50, none lasted more than a decade.

Other Minerals

As well as gold, other minerals have been won from time to time in the Murchison. We have already noted the small copper and lead mines which operated in the 1800s. Tin, talc, tantalite, minerals sands, iron ore and natural gas have each been contributors to the output of the region and discoveries of uranium at Yeelirrie, nickel at Mount Keith and copper-zinc at Golden Grove.

Iron ore warrants a mention because of its pioneering role. After the lifting in 1960 of the embargo on the export of iron ore, several companies began to search for iron ore deposits in Western Australia. The first to export iron ore commercially to Japan was Western Mining Corporation Ltd through Geraldton, in March 1966, from Koolanooka, north of Morawa. It was a very small deposit compared with the vast developments in the Pilbara, but it was significant as the initial entry into the Japanese market.

Much of the Murchison and nearby districts was explored and sparsely settled by pastoralists before the gold rushes. Their presence aided the movement of prospectors through the area by providing a network of tracks and some watering places. They were also a source of food and

Using an air-driven bogger at Hill 50 gold mine, early 1980s. The mine was named to commemorate an Australian military objective in France during the First World War.

Iron ore mining at Koolanooka. Source of the first iron ore exported under long-term contract to Japan, Koolanooka began production in 1966. Ore was shipped through Geraldton.

Rubber-tyred diesel-powered equipment at Golden Crown mine. Such equipment as this enabled a much more flexible approach to mining than rail-mounted equipment.

supplies. Some gold discoveries were made by pastoralists, their families or station workers. Landholders may not have been pleased about the intrusion of prospectors on their land, but they were to benefit from the improved services which followed the mining developments. Roads, railways, telegraphs and ports, although slow to be developed, were obviously advantageous to all and the proximity of a mining town provided facilities that pastoralists and their families would not otherwise have enjoyed.

Western Australian gold mines were a steady source of trade to machinery merchants and manufacturers.

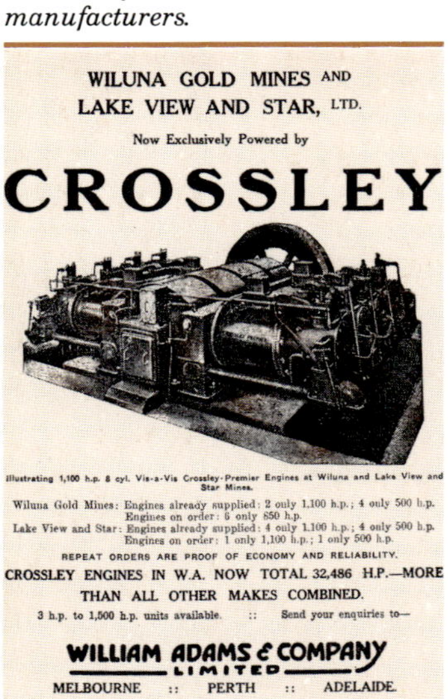

Revival and Prosperity

Annual production in the greater Murchison area now exceeds 1,000,000 ounces. The main producing areas are: Meekatharra (240,000 ounces), Lawlers (220,000 ounces), Mount Magnet (140,000 ounces), Wiluna (130,000 ounces), Day Dawn/Cue (100,000 ounces), Sandstone (95,000 ounces), Peak Hill (60,000 ounces), and Yalgoo (60,000 ounces).

Colourful Characters

It might be of interest to refer to some more of the events and stories of the past – the romance of the region.

Russian Jack, whose real name was Ivan Fredericks, was often in clashes with the local law. D.W. De Havelland in *Gold and Ghosts* records how 'outside the Nullagine Police Station lay a large solid log, with three or four ring bolts fastened onto it. To these, prisoners were attached. Russian Jack, gaoled for his drinking offences, would awaken, come to his senses, and suffering from the DTs, up-end the log, carry it into the nearby hotel and sink one or two refreshing ales. Having controlled his immediate urge, Jack would return himself and the log to the appropriate position outside the Police Station'.

Russian Jack's strength was immense. In later years he was in work on a station as a general hand when he was sacked after bending a four centimetre octagonal crowbar over his knees.

Carr Boyd is a well known name throughout the Western Australian goldfields, principally because he was one of its greatest story tellers. Malcolm Uren says of him that, 'many of the stories were true; some of them were based on fact and garnished with imagination; others were the spontaneous product of a born raconteur'. On one occasion when responding to a toast by Sir John Forrest, Carr Boyd remarked, 'There is so much gold in the Kimberley that its value would depreciate'.

The spirit of goldfields generosity is typified by the case of a woman who once arrived in Sandstone with three famished children and the body of her husband crammed into the back of a wobbly old cart drawn by a feeble horse. The deceased, who was wrapped in bags, was given a decent burial while the woman and her children were provided with clothes and sent off to the coast on the proceeds of the sale of the wrecked cart and wretched horse which were auctioned, returned and re-auctioned several times until more than £100 had been raised.

The story is told by Bob Sheppard, in *Barb of the Spear*, of a prospector arriving at the Pearly Gates only to be told by Saint Peter that there was no room left as heaven was already full of prospectors. As he sadly wandered away he suddenly had an inspiration, cupped his hands, and yelled: 'Gold, gold, I've struck gold'! The gates of Heaven burst open and hundreds of prospectors rushed out. He then went up to Saint Peter who now beckoned him in. As he went in he said to Saint Peter: 'Leave the gate open, will you mate, there might be some truth in that rumour'.

Carr Boyd, one of the great story-tellers of Western Australian gold mining.

Future Outlook

The recent revival of the gold mining industry occurred not only because of the better price of gold but also because of technological advances in exploration, mining and mineral processing.

The history of nearly 150 years of mining in the Murchison spans three quarters of the history of European settlement of Australia. The contribution to the development of Western Australia and Australia has been irreplaceable.

While ups and downs in the level of activity are a characteristic of the industry, the important feature of the Murchison is that it is a highly mineralised part of Australia. More exploration is likely to result in more mineral finds. I have no doubt that mining will remain an important activity in the Murchison for decades and perhaps centuries to come.

I am indebted to Gilbert Ralph, MBE, and Bob Watchorn of WMC Mt Magnet for their invaluable assistance in the preparation of this paper, which was presented in Meekatharra on 31 May 1990.

Iron Ore, Oil and Gas from the Pilbara

It has become commonplace to say that Australia's future lies with Asia – that we should be more active in developing links with that part of the world, so that we will participate in the rapid growth taking place there and so that the Australian community will learn more about its Asian neighbours.

The Pilbara has been forging those links for many years. Perhaps more than any other region in Australia it has benefited from and contributed to the 'economic miracles' of Japan and, more recently, Taiwan and South Korea.

Early Days: Pearls and Gold

Some of the region's links with Asia go back to the days before gold fever gripped the area. In the days when pearling was a more important activity than grazing, many Asians – Malays, Koepangers and, later, Japanese – worked along the coast as divers.

The explorers F. T. Gregory in 1858 and 1861, Warbuton in 1873, Ernest Giles in 1876 and Alexander Forrest in 1879 traversed and criss-crossed the Pilbara. It appears that Gregory passed within 15 kilometres of Mount Tom Price, and Giles came to about 15 kilometres from Mount Newman before retracing his steps. F. T. Gregory in the notes that accompany his description of his 1861 journey through the Pilbara records: 'Minerals, I was unable to discover any traces. Except iron'.

High technology mining had not yet reached the Pilbara when this photograph was taken in the Marble Bar area about 1960. It shows a yandy dish being used to concentrate cassiterite.

In 1888-89 H. P. Woodward, the Government Geologist recorded: 'This is essentially an iron country ... there is enough to supply the whole world ...' In 1892 he wrote a *Mining Handbook* and stated: 'Iron is the commonest occurrence in the colony ... but the ores are of no value, it is useless to trouble about them'.

A gold rush to the Pilbara began on Christmas Day 1887, when young Jimmy Withnell picked up a stone to throw at a crow, and found it flecked with gold.

The Pilbara Goldfield was officially declared in 1888 – the same year the Perth Stock Exchange opened for business – with many of the diggers coming from the Kimberley, where the rewards had been rather poor. The Pilbara produced a good deal of alluvial and reef gold. Statistics on the finds are flimsy, but the Pilbara is said to have produced more 50-ounce nuggets than any other field in Western Australia.

When the more accessible gold had been taken, the diggers moved on. News of discoveries at Marble Bar and Nullagine drew men inland to the hottest part of the continent, where apparently even a spell of 160 successive days of century heat did not dull the appetite for gold.

As we know only too well, no boom lasts forever. The activity which accompanied the gold rush soon ended. Isolation, a harsh climate, and

This tin-tantalite mine at Wodgina was once owned by Dr Deborah Murphy the "Tantalite Queen."

lack of water – the factors which have been so important in shaping Australia's history – played their customary roles in constraining development for many years.

A Long Pause

Although gold mining continued on a small scale, high operating costs eventually forced most mines to close. Other mining developments followed in the wake of gold. Tin was first discovered on the Shaw River by a man named Eley in 1890, but it was the discoveries at the turn of the century which really sparked off the tin mining industry. Production peaked in 1907 and dwindled to almost nothing by the thirties.

Alluvial tin was found at Wodgina, by Ogilvy and McArthur in 1903. Since then, Wodgina has produced almost every kind of mineral at some stage in its history, the most important being tin and tantalite. One of the mine's early owners, Dr Deborah Buller Murphy, who took up an interest in it in 1923, became widely known as the 'Tantalite Queen'.

Copper was first discovered near Roebourne in 1872, and Whim Creek was Western Australia's premier copper mine between 1909 and 1914. The Pilbara also yielded white and blue asbestos in the 1930s and 1940s. Numerous small deposits of manganese were mined, together with some lead and zinc.

Other developments followed. As tin mining at Marble Bar became more important, demand for a rail link between Marble Bar and Port Hedland grew. This was approved in 1907, thereby catalysing another development – the construction of a new jetty at Hedland, which opened in 1909. The rail line was officially opened in July 1911.

After that, things more or less stalled. Mineral production in the Pilbara actually declined between 1900 and 1940.

The Second Wave

The Second World War stimulated demand for minerals such as tantalite and manganese. At one stage during the war, tantalite was needed so urgently that 22 tonnes were taken from the Pilbara to the United States by air. About the same time an urgent market developed for beryl, and a 900 tonne shipment to the United States was the biggest consignment of beryl in the world. But after the war interest in these minerals quickly fell away again. The 1950s saw some growth, with a revival of tin mining and development of lead and zinc mines in the Ragged Hills area. Mining of asbestos – not in those days recognised as harmful – continued, reaching peak production in 1962. West Australian Petroleum Ltd (WAPET) began searching for oil in the region in 1952. They found some inland from Exmouth Gulf, but not in commercial quantities. On the whole, things were quiet.

Successive state governments talked about developing the vast, empty north, but in reality little changed. By 1954 the whole of the Pilbara region – an area about twice the size of Victoria – had a population of only 2,600. Port Hedland was home to about 700 of them and 'tottered along on the edge of collapse for decades'.

Stan Hilditch, discoverer of iron ore at Mount Whaleback, with his wife.

Little by little, technology provided the liberating force for Australia's outback areas from extreme isolation. Some of those who recognised the role such developments as air travel could play went on to become legendary figures.

The Reverend John Flynn, who founded the Flying Doctor Service, was instrumental in setting up the nursing home which opened in Port Hedland in 1915. And Charles Smith – later Sir Charles Kingsford-Smith – was a familiar face in these parts in the 1920s, flying mail and passengers in and out. Prior to that, monthly visits of the steamer were the main contact with the outside world. 'Smithy,' and the other 'fly-boys', landed on the Port Hedland racecourse, which served as an aerodrome until 1940.

There were other changes. At Port Hedland for instance, adequate water supply was finally secured in 1953, by piping water from the Turner River. Until then few trees could be planted and the available water had a high mineral and saline content. Water pipes clogged easily and pressure was poor, particularly, it seems, in the hotels. One early MMA (MacRobertson Miller Airlines, also known affectionately as Mickey Mouse Airways) hostess recalled:

> In all other towns where we 'overnighted', the usual invitation from the locals was, 'Come and have a drink' but in Port Hedland the invitation was 'Come round and have a bath'.

The Iron Ore Era

For all the improvements in communications and transportation, the Pilbara remained remote until the early 1960s. Many of these developments, particularly in shipping and rail technologies, were later to

Mining iron ore at Paraburdoo, 1970s.

play a pivotal role in transforming the Pilbara, but some entrenched ideas had to change before their impact could be felt.

Fears about the scarcity of mineral resources seem to crop up on a regular basis. Most of you will recall the dire predictions in the Club of Rome's report *The Limits to Growth* in 1972 – 'known' reserves of tin were only enough to last another 15 years and so on. Forty years prior to that things were even more alarming; one study reported in 1929 that known reserves of tin would only last 10 years!

What these reports failed to take into account is that mining people do not 'prospect for eternity'. Known reserves never represent the ultimate reserves in any meaningful way, but rather those reserves which it has been economically attractive to locate.

In the post-war period, successive Australian Governments failed to grasp this. Exports of iron ore were banned in July 1938 for security reasons. After the Second World War the ban remained through fear of shortages.

Throughout the 1940s and 1950s this was widely accepted. By retaining the ban on exports the Federal Government ensured that there would be little or no exploration. The Western Australian Government did its bit too, by refusing to grant prospectors or companies titles to explore for iron ore, and by awarding mining rights not to the discoverer but to those who agreed to bring manufacturing industries to the West.

The impact of this is now well known. When Stan Hilditch realised in 1957 that he stood atop a mountain of iron ore – later known as Mount Whaleback – he revealed his discovery only to his wife. The couple kept it a secret until 1961, when the embargo on iron ore exports was lifted.

Loading iron ore at Hamersley's Mount Tom Price mine.

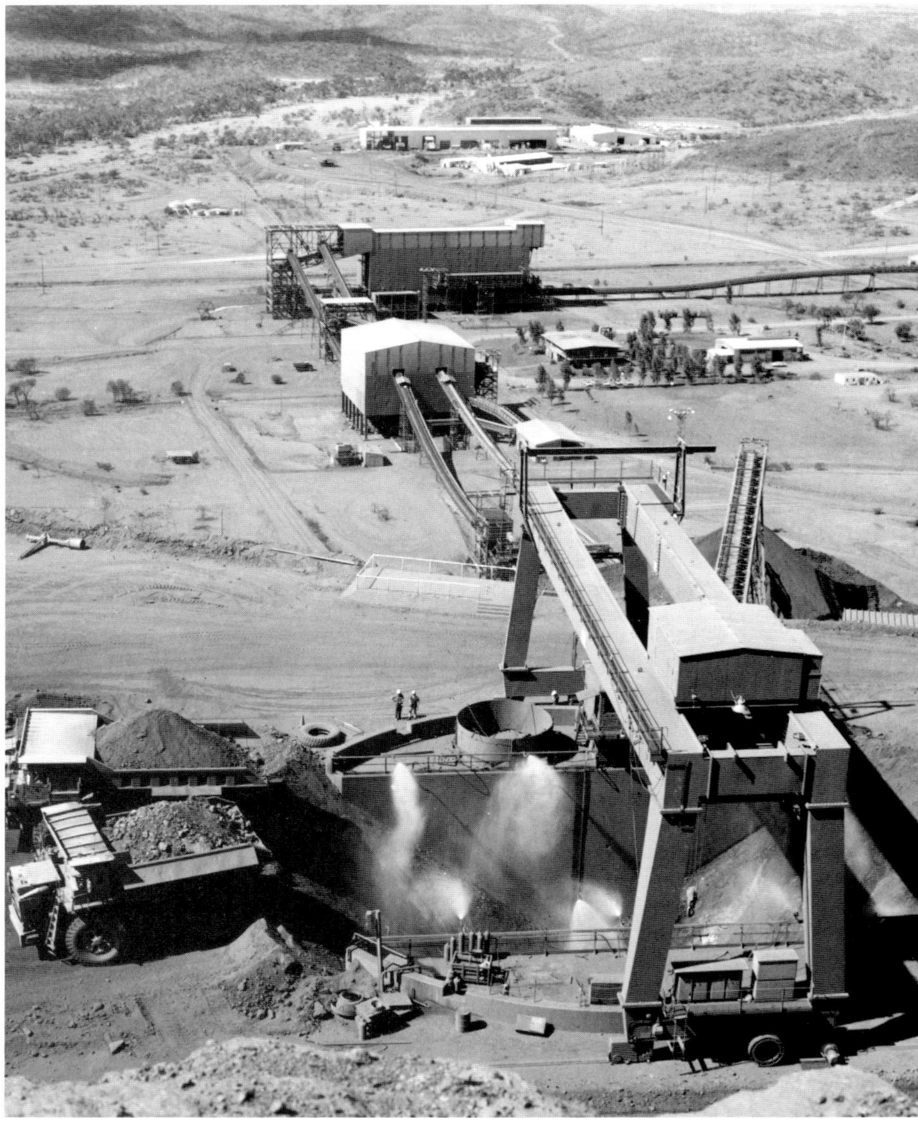

Crushing and screening plant at Hamersley mine at Paraburdoo, 1970s.

Late in 1960, Australia's low foreign currency reserves prompted the Federal Government to modify the ban on iron ore exports. The 'glow of Japanese steelworks' could be seen even in Canberra, and the temptation to export iron ore to Japan could not be resisted any longer. But it was not until the next year that the Western Australian Government started to grant prospecting and mineral leases.

An exploration and export boom resulted. For the Pilbara, it marked the beginning of a new era. Stan Hilditch's secret discovery came to light, together with that of Lang Hancock, a pastoralist who, according to what is now local legend, spotted in 1952 an outcrop of iron ore when flying over the Hamersley Ranges. Together with the ore bodies discovered along the Robe River – by BHP geologists and on the upper reaches by Garrick Agnew – these vast reserves became known as the Hamersley Iron Province, estimated to total over 30 billion tonnes. At last technical developments, notably in rail and shipping technologies, would be put to good use in the region.

The first iron ore mine to come into production in the Pilbara was Mount Goldsworthy. The first shipments of ore left for Japan in June 1966. Regular shipments over the next seven years earned a then staggering $144,000,000. This would be, however, dwarfed by what would be won from other Pilbara mines.

Development of the Hamersley Operations

Following the lifting of the embargo on iron ore exports, other companies had exploration teams in the Pilbara. Among them was The Rio Tinto Mining Company of Australia Ltd (RTMA), introduced to the region by Lang Hancock while the embargo still applied. Hancock was able to show the company large deposits of limonitic ore and near his homestead some small deposits of haematite. The RTMA team was directed by Ben (later Sir Ben) Dickinson, with field geologist including R. S. Matheson, Harold Jensen, Dr Bruno Campana and Frank Hughes.

Consolidated Zinc Pty Ltd (CZP) geologists joined the search in March 1962. This was in anticipation of the merger between the two companies (RTMA and CZP) to form Conzinc Riotinto of Australia Ltd (CRA). The team was then led by Haddon King, and in the field by Frank Hughes.

Train loaded with almost 15,000 tonnes of iron ore sets out from the mine at Mount Newman to begin the 426 kilometre journey to the coast at Port Hedland, about 1970.

Its work led to the recognition in September 1962 by Bill Burns and Ian Whitcher of an enormous haematite deposit, destined to be called Mount Tom Price, 40 kilometres outside the granted Temporary Reserves.

Although Lang Hancock did not discover the major haematite deposits that became the basis of Pilbara iron ore mining, he certainly played a vital role in bringing it to the attention of A. J. Keast of RTMA.

Systematic drilling revealed at least 500,000,000 tons of rich ore and many hundreds of millions of tons of lower grade ore – in all more ore than was known to exist in Australia only three years earlier.

Laying the track for the railway to carry iron ore from the mine at Mount Newman 426 kilometres to the coast at Port Hedland.

No time was wasted in developing the ore bodies. Construction on the project began early in 1965 and by August 1966 the mine at Mount Tom Price, the port and loading facilities at Dampier and of the 386-kilometre railroad linking mine to port were ready for operation. Twelve months later the first shipment of ore left for Japan.

Hamersley Iron opened a second mine and established the nearby township of Paraburdoo in 1971. In the late 1980s, a third mine, Channar, came into production, consolidating Hamersley's position as one of the world's largest iron ore producers and suppliers and the largest single supplier to Japan and North East Asia.

Development of the Mount Newman Project

In 1963 AMAX was granted an option over Hilditch's discovery. Many companies declined the chance to become involved in mining the ore, believing it to be too far from the coast. By the time a joint venture was finally launched in 1967, AMAX had reduced its stake. BHP had joined the consortium and became managing partner.

Development of the Mount Whaleback ore body began in 1967. In contrast to early mining activity in the Pilbara, everything about the project was on a colossal scale. The mine was to become the largest open-cut iron ore mine in the world. The town of Newman was established the following year and soon had twice as many people as the whole of the Pilbara in the 1950s. The tidal port at Hedland – chosen as the port from which the ore would be shipped – was transformed into one of the biggest export ports in Australia. Once restricted to vessels of 5,000 tonnes, it became capable of handling 260,000 tonne ore carriers.

William Walkley, Managing Director of Ampol, at the spudding in of Ampol Exploration's well at Rough Range, in the Pilbara. The well struck oil, but not in payable quantities.

A 426 kilometre railroad, with the heaviest rail then used in Australia, was built between the mine and port. It remains one of the longest privately operated railroad in the world, and its efficient operation in large part underpins the viability of mining so far inland.

In April 1969, the first shipment of Newman iron ore was loaded at Port Hedland. Originally the project was planned to ship 5,000,000 tonnes of ore per year. The first year saw 6,600,000 tonnes shipped, and production rose dramatically after that.

Not surprisingly, new 'legends' have emerged from all this activity. Construction of the railroad from Mount Newman to Port Hedland involved up to 1,300 men and required the development of special equipment. Several records were established, including some for track-laying and others for consumption of beer. It would seem Bishop Witt was correct

Loading iron ore for export by Robe River Mining Company, Port Walcott.

when, some time later, escorting the teetotal Archbishop of Sydney on a visit to the area in 1971, he was able to assure the Most Reverend Marcus Loane that Newman did not have a drinking problem – 'you can usually get served pretty quickly'.

A number of milestones have been passed since the first shipment of iron ore left for Japan. Notable amongst them was the shipment of Mount Newman's 500,000,000th tonne of ore in 1988. Along the way the Marra Mamba orebody was developed, the beneficiation plant at Newman commissioned, and the Joint Venture partnership changed.

The oil 'shocks' of the '70s brought the post-war era of continuous world economic growth to an end and ushered in a period of uncertainty and readjustment. This was true not just at Newman, but for the other iron ore producers in the region. As John McIlwraith observed in *The First 500 Million*: 'The 1980s could be described as the decade in which the Pilbara industry lost its innocence'. Happily, by the end of the 1980s the outlook was starting to improve – the iron ore market strengthened the prices increased after six years of decline.

Petroleum exploration

Petroleum exploration in the Pilbara began in 1952, when West Australian Petroleum Pty Ltd (WAPET) took up permits covering large areas both on and off-shore. Remarkably WAPET struck oil with their first exploration well, Rough Range No. 1. Although not commercial, this find catalysed a boom in exploration and during the next decade more than 70 new wildcat wells were drilled.

Exploration of Barrow Island proved more successful for WAPET. The first shipment of crude ore left the small, previously uninhabited island 96 kilometres north of Onslow, on 25 April 1967.

A small Victorian company, Woodside (Lakes Entrance) Oil N.L., was interested in the underwater area north of the WAPET leases, the North West Shelf, a vast area as big as the British Isles. Woodside's interest – which was based on a report by their chief geologist, Dr Nicholas Boutakoff, – proved to be well founded. The company was granted exploration permits in 1963, and sought partners to help finance what

Rough Range No. 1.
First discovery of oil.

became, in true Pilbara tradition, a massive project – in fact, the largest resource project in Australia. The Joint Venture was formed in 1964 and the first well drilled in 1967. The well was dry, but a second one did find some oil, although not in commercial quantities. This pattern continued for the next four years.

Gas Finds

The disappointments were compensated for when a large gas/condensate field was discovered in an area named North Rankin late in 1971. This was soon followed by further finds at Goodwyn and Angel. The three fields are about 130 kilometres off the coast in water as deep as 130 metres.

Drilling rig at Barrow No. 1 oil well, 1964.

Major obstacles stood in the way of developing these fields. The cost involved was huge and could not be justified simply to supply the Western Australian market alone. Developing an export market was crucial. Japan and the United States were the only markets of suitable scale in the Pacific region.

Technical problems posed further challenges. In the days when decisions were being taken on the viability of the project, deep offshore drilling technology was in its infancy and no company was producing gas at such depths. And, of course, the factors which hampered growth in the region for so long – isolation and a harsh climate – also had to be overcome.

So the hurdles were many. Even so, the project went ahead. It has involved massive investment in both on and off-shore facilities and transformed the Burrup Peninsula. The town of Karratha boomed. Once reliant upon the iron ore and salt industries, its population almost doubled once the North West Shelf project began and its facilities improved enormously.

Gas sales to the State Energy Commission of Western Australia began in 1984. By 1988 the North West Shelf was supplying 95 per cent of the State's industries and 200,000 households with natural gas. The first shipments of LNG to Japan, in 1989, reaffirmed the Pilbara's close links with that country.

North Rankin 'A' gas and condensate production platform on the North West Shelf, 134 kilometres north-west of Dampier.

As with the minerals industry in the Pilbara, technological developments have been crucial in the success of this project. Heavy lift derrick barges and large capacity launch barges, developed for work in the North Sea, have been particularly important. Australian ingenuity has also come to the fore, in developments such as the underwater plough and the Dawson Schimitar.

The economic impact of the project stretches well beyond the Pilbara. The State Government estimates that $4.4 billion has found its way into the Western Australian economy since the project's inception and it has stimulated growth in other parts of the State. A liquefied petroleum gas (LPG) extraction plant was built in Kwinana and a number of other projects planning to use gas are committed or awaiting approval.

In 1990, the story of the North West Shelf continued, the next chapter being construction of the Goodwyn A platform. The total investment in the project, when completed, will be $12 billion.

Harvesting solar salt from the Dampier field which was established in 1967 and is now exporting about four million tonnes annually.

Other Major Developments

There has been a revival of interest in other minerals too. Once again gold is of importance to the region. Technical developments have enabled some of the difficulties of old to be overcome. The Telfer mine, which came into production in 1977, was until 1988 the largest producer in the State.

Production of tin concentrate continues. It is well down from peak production of the 1960s, but the region still accounts for about 40 per cent of the State's output. Wodgina started a new lease of life in 1990 when a multi-million dollar tantalite processing plant was brought on stream. Mining of manganese has recommenced, the first shipment since the 1960s leaving Port Hedland in 1990. Technical developments have also brought copper production in the Throssell Ranges within reach.

Salt is another significant industry in the region. Dampier Salt Ltd, at Dampier, and Leslie Salt Ltd, at Port Hedland, are among the world's largest solar salt producers. Construction work on extensions to Dampier Salt's operations increased capacity substantially, taking the company's annual production capacity to 4,100,000 tonnes. Dampier Salt's product carries the distinction of being regarded by the Japanese as the highest quality of all salt imports.

Future Outlook

It was recognised early that the Pilbara was a great storehouse of mineral wealth. Today, the region accounts for some 40 per cent of Western Australia's mineral output. With huge iron ore and gas reserves, the Pilbara will continue to dominate the State's mineral production in the foreseeable future.

By 1990, several recent developments ensured this would be the case, with a new LNG phase of the North West Shelf Project underway and completion of Fractionation Train 3 and the Goodwyn Platform scheduled for mid-1993. Results from Angel 4 had raised confidence about the ultimate size of the Angel field. Further drilling in the Shelf area had also reinforced the region's growing potential as an oil province.

Production of iron ore had begun at Channar. The new mine was a joint venture between Hamersley Iron and China Metallurgical Import and Export Corporation, with all output exported to China. Mount Goldworthy's extension programme, which included construction of a beneficiation plant, would ensure that the project which yielded the first shipment of ore from the Pilbara would continue.

Technological developments will doubtless bring more mineral reserves within economic reach and may encourage completely new finds. The viability of mineral processing plants will no doubt be explored vigorously in the future.

The strong ties with Japan and other parts of Asia will remain of great consequence to the region, and to Australia. The remarkable changes we have seen recently in the Soviet Union and Eastern Europe suggest too that in the future, minerals of the Pilbara may also have a role to play in developments in other parts of the world.

Liquified natural gas (LNG) carrier Northwest Sanderling, built to carry LNG from the North West Shelf to Japan.

I am indebted to Vicki Lukins for her major contribution and Geoff Blackman and John Crawley of BHP Minerals, John Dew and Gilbert Ralph, MBE, for their assistance in preparation of this paper which, with variations, was presented at Newman on 28 May, Tom Price on 29 May and Karratha on 30 May 1990.

Onshore processing facilities for LNG on Burrup Peninsula. The North West Shelf Project is the largest resources development in Australia's history.

Coal, Mineral Sands and Bauxite from the South-West

The pioneers who came to settle the Swan River colony in 1829 had little interest in minerals. Not until 1839 did the Colonial Government offer 1,000 acres of land to anyone who discovered a payable coalfield. In 1846 A. C. Gregory, a government surveyor, claimed the reward for a seam of coal on the banks of the Irwin River near Mingenew.

Interest in Western Australia's mineral resources was stimulated by the prosperity created by the copper discoveries in South Australia. In 1846, B. Urban Vigors, a Perth lawyer, formed the Western Australian Mining Company, which bought a lead prospect near Kelmscott. Dr Von Sommer, a German geologist, came from South Australia to manage the mine, but it failed to produce enough lead and was closed in 1848. At about the same time, A. C. Gregory discovered galena on the Murchison River, and this led to the successful Geraldine lead mine.

By 1854 Western Australia was in economic depression. Prompted by the Eastern States gold discoveries, interested people raised £357 as a reward for the discovery of payable gold. A few claims were made but nothing significant came of them.

H. Y. L. Brown, who was appointed Government Geologist in 1870, reported small quantities of gold at several places in south-west Western Australia, notably Kelmscott, Toodyay, Boyup, Serpentine, Stirling Range, and Phillips River. According to M. K. Quartermaine and E. McGowan in *Mining in Western Australia*:

> Other discoveries led to the formation of the first two gold-mining companies in the colony. The first, at Cardup in early 1874, petered out before a hole was dug; the other, the Standard Gold Mining Company, was floated in Albany on a capital of £1,500 and became the first company in the colony to attempt to exploit a gold discovery.

The mine, near Kendenup, only lasted a few years. When gold rushes did come to Western Australia, they began in the Kimberleys in 1885 and carried through to Kalgoorlie in 1893. They attracted enormous interest and little further prospecting for gold was done in the south-west.

Title page from Government Geologist Harry P. Woodward's 1894 Report on the Collie Coalfield.

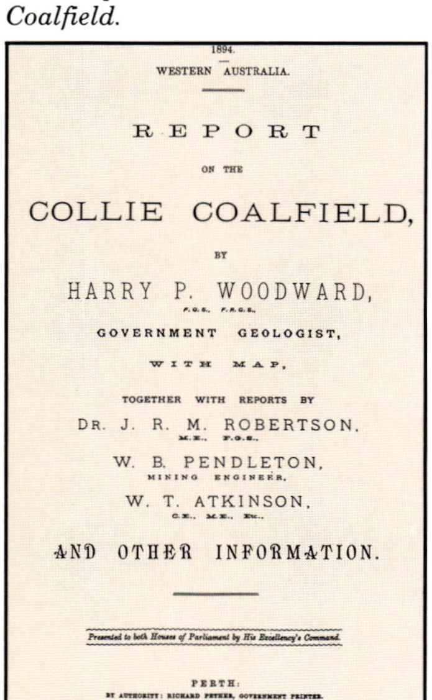
1894.

WESTERN AUSTRALIA.

REPORT

ON THE

COLLIE COALFIELD,

BY

HARRY P. WOODWARD,
F.G.S., F.R.G.S.,

GOVERNMENT GEOLOGIST,

WITH MAP,

TOGETHER WITH REPORTS BY

DR. J. R. M. ROBERTSON,
M.E., F.G.S.,

W. B. PENDLETON,
MINING ENGINEER,

W. T. ATKINSON,
C.E., M.E., Etc.,

AND OTHER INFORMATION.

Presented to both Houses of Parliament by His Excellency's Command.

PERTH:
BY AUTHORITY: RICHARD PETHER, GOVERNMENT PRINTER.

No. 9. 1894.

Coal

The south-west did, however, benefit from those gold rushes, for it had another, less-glamorous mineral. The gold rushes brought people, development and optimism. The demand for goods and services grew rapidly, including the need for coal for the State's burgeoning railway system, which at first used imported coal. In 1887 the Government announced a reward for the discovery of a payable coalfield, the Irwin River find having failed to live up to expectations.

Catherine Stedman in *100 Years of Collie Coal*, says that 'the discovery of coal on the Collie River remains to this day a story of mystery and intrigue. Perhaps the strongest claim to be the initial discoverer belongs to a humble stockman, George Marsh', a shepherd on Arthur Perren's pastoral lease. He claimed in October 1883 that, to his surprise, the stones he was using to support his cooking utensils caught alight. A number of attempts failed to define the extent of the coal seams.

In 1890 the Government Geologist, Harry Woodward, reported favourably on the deposit and in 1893 Dr Robertson was engaged to test it.

Knowledge of the field was advanced sufficiently for the Premier, Sir John Forrest, KCMB, GCMG, to announce in 1894 that the deposit would be developed. By the end of the century production exceeded 50,000 tonnes a year, mostly from the Wallsend mine.

Amalgamated

A succession of companies produced coal from the Collie field, beginning with Coalfields West, Coal and Fireclay Ltd in 1897. Several companies merged in the 1920s to form Amalgamated Collieries Ltd, which dominated the field for 40 years. Amalgamated struggled through the Depression, was caught unawares by the sudden increased demand during the Second World War, and soon after the war developed open-cut mines.

During the early 1950s the Government, which to some extent controlled the field, insisted on greater development of underground mines. In 1955 it abandoned the cost-plus system of pricing in favour of tendering and the State Electricity Commission of Western Australia, which was by far the field's biggest customer, pressed Amalgamated to reduce its price. It held out, and in December 1960 the contracts were let to Griffin Coal Mining Pty Ltd and Western Collieries Ltd. Amalgamated was completely excluded and forced to retrench its 700 workers.

Early days at Griffin Syndicate mine, Collie

Griffin

The Griffin Syndicate was formed in 1923. John Ewing, MLC, was a member and provided the political influence needed to 'fight the Amalgamated Collieries monopoly', as the syndicate struggled to win contracts to supply coal for the Government railways and power stations. Dr Hyman Herman, engaged as Royal Commissioner into the industry, found in 1931 that, 'Griffin was entitled to a share of the Government order at a price proportionate to the quality of its coal'. Dr Herman, who was also associated with the development of Victoria's vast brown coal deposits, was awarded the Institute Medal for 1938.

The Griffin Coal Mining Company, which was registered in 1927, then received eight per cent of Government business and from this base it steadily grew, but not without numerous setbacks during the 1950s, when its contracts were cut and a fire closed the Wyvern Colliery. Development of the Muja open-cut and winning the contract to supply the nearby State Electricity Commission Muja power station were important events for the Company.

Western

The other major company on the Collie field, Western Collieries Ltd, had its beginnings in Kalgoorlie when a group of mine managers, concerned by the dwindling supplies of firewood on the Goldfields, formed the Goldfields Coal Syndicate in 1947 and pegged two coal prospecting areas near Collie. These proved to contain sufficient reserves to form Western Collieries, and by December 1950 the company was producing from an open-cut. Both underground and open-cut mines were developed, and Government and private contracts enabled the company to grow.

Western Collieries Muja open-cut coal mine and the State-owned power station.

Mineral Sands

Black sands along the beaches of the south-west coast of Western Australia had been observed for many years but their value was not recognised until 1949, when the Cable brothers extracted 158 tonnes of mineral sands from beach sands at Cheyne Bay.

Demand for ilmenite as a source of titanium oxide for pigments prompted further development of the mineral sands deposits, with Cable Sands Pty Ltd starting operations at Koombana Bay in 1956, Western Titanium Ltd at Capel South in 1957, Westralian Sands Ltd at Yoganup in 1959, Ilmenite Minerals Pty Ltd at Wonnerup in 1959 and Westralian Mineral Sands Pty Ltd at Capel North in 1964. Other sites were later developed by some of these companies.

From January 1950, the Federal Government prohibited the export of crude mixed concentrates of mineral sands, and this caused the industry to develop improved techniques for separating ilmenite, zircon, rutile and other components.

One contributor to this technology was Ernst Reichert, a German electrical engineer, who was working for Siemens in Iran when the Second World War broke out. Taken prisoner and interned in Australia, he applied his inventive mind to electrostatic separation of minerals,

Loading mineral sands into rail waggons, TIWEST Joint Venture, Muchea.

which he developed in 1946 when working in the beach sands industry at Southport, in Queensland. He later developed the Reichert Cone, which found wide use in the industry throughout Australia.

Until 1948, (when Tioxide Australia Pty Ltd built its plant at Burnie, Tasmania) all white pigment used in Australia was imported. The growth of the mineral sands business in Western Australia prompted the building of the Laporte Titanium (Australia) Ltd plant in Bunbury in 1964. These and other such plants became important contributors to downstream processing and import substitution.

Western Titanium built a beneficiation plant at Capel in 1968 to produce synthetic rutile, which increases titanium oxide in the product from 54 to 92 per cent. This was the first commercial synthetic rutile plant in the world and it incorporated many innovative features developed by Bill Hockin, Western Titanium's Technical Director.

The mineral sand deposits in the Eneabba area, north of Perth, were discovered in 1970, and steadily developed. The northern deposits differ from those in the south-west in that they contain substantial quantities of rutile and zircon, as well as ilmenite. AMC Mineral Sands Ltd, the world's largest mineral sands producer, operates in the Eneabba area and in the south-west.

The beach sands industry, like other sectors of the minerals industry, has been subject to wide variations in prices. In 1957, as the Western Australian industry was beginning to establish itself, rutile was £70 per ton, but by 1961 it was down to £30 per ton. Fortunately the price recovered, and the industry not only survived but expanded to become a significant part of the minerals industry.

In the early 1990s, Australia produced around 50 per cent of the western world's rutile and ilmenite, 60 per cent of its zircon, and 40 per cent of its monazite, with Western Australia contributing the majority of Australia's production. In 1989 the value of exports was about

Bauxite on its way from the Jarrahdale mine to the Alcoa alumina refinery at Kwinana.

$800,000,000, and Australia-wide the industry employed 3,250 people. Strong demand and prices in the late 1980s encouraged further expansion of the industry, particularly in Western Australia, where the reserves are large and the Government had been less influenced than in Eastern Australia by the objections of a vocal minority of opponents.

Developments in the 1980s included the TIWEST Joint Venture, which combined mining and concentrating at Cooljarloo, mineral separation and a synthetic rutile plant near Muchea, and a pigment plant at Kwinana. Further secondary processing facilities were established at Kemberton and Rockingham. Four synthetic rutile plants were built in Western Australia to convert ilmenite to synthetic rutile, with a combined annual output of about 270,000 tonnes.

A feature of the mineral sands operations in Western Australia is the very successful rehabilitation of mined areas. This can be seen by visitors who are often unaware, until the fact is brought to their notice, that some of the beaches and large areas of grazing land have previously been mined and rehabilitated. Native species were replanted and some farm lands enhanced. An area of wetlands, developed by AMC as a Nature Reserve, is an added attraction to the region and conducted tours are popular.

Tin – Tantalite – Lithium

According to Harry P. Woodward's 1908 report, the discovery of tin in the Greenbushes area is attributed to E. T. Hardman, Government Geologist, who noticed stanniferous material from a well in 1881 or 1882. He advised D. W. Stinton, a local kangaroo hunter, to prospect the area for tin. Stinton did so, found tin and pegged the Bunbury lease in about 1888. Small-scale development took place and production rose to 783 tons in 1906, then steadily declined to less than 100 tons during the 1920s and almost ceased during the 1930s.

Tin treatment plant, Greenbushes.

Many different methods were used to recover the alluvial tin, including puddling, sluicing, bucket dredging and shaft sinking on the lode system which, although discovered in 1893, was not worked extensively until after 1937. The first bucket dredge was tried in 1938-39 but it was not a success.

There was a revival of interest in the field in the 1950s but it was not until after the area had been drilled by Aberfoyle Tin N.L. in 1961-64 that Greenbushes Tin N.L., formed in 1964, installed a second-hand bucket dredge in 1966 and production increased markedly. The dredge was salvaged from the Amphitheatre operations of Central Victoria Gold Dredging Co. N.L.

At first the dredging was successful, but it ran into problems in the late 1960s and the emphasis shifted to open-cut mining of large kaolinised pegmatites. These were treated in the 'dredge', which was incorporated in a static plant in 1970. Annual production reached 1,500 tonnes but later fell to less than 500 tonnes of contained tin, with the Greenbushes tin smelter the only one operating in Australia.

The laboratory at Kalamunda which was set up by Western Mining's subsidiary Western Aluminium N.L. to test bauxite samples from the Darling Ranges.

Although Greenbushes was known mainly as a tin company, it had since the mid 1970s supplied approximately 15 per cent of the western world's tantalum requirements, and in 1990 was the largest single tantalum producer in the world. In more recent times a solvent extraction plant was installed to recover tantalum pentoxide from concentrates produced at the Greenbushes mine. An associated company, Lithium Australia Ltd, mined and concentrated spodumene from the Greenbushes pegmatite and was a major source of lithium in Australia.

Greenbushes Ltd, which became part of the Gwalia Minerals N.L. group of companies, accounted for more than half of Western Australia's tin output which, while only a small percentage of Australia's production, had, together with tantalum and lithium products, contributed to Western Australia's mineral exports.

Alumina

The discovery, growth, and development of the bauxite-alumina industry in south-west Western Australia has been marked by numerous milestones.

Its two producers, Alcoa of Australia Ltd and Worsley Alumina Pty Ltd, in 1990 produced about 23 per cent of the western world's alumina. This brought very substantial export income to Australia and provided significant direct and indirect employment for Western Australians in one of the State's more pleasant regions.

Although the bauxitic nature of the laterites in the Darling Range was known from the early part of the twentieth century, it was not thought to be of economic grade until examined more closely by Western Mining Corporation in the late 1950s.

In 1957 WMC set up a laboratory at Kalamunda to analyse samples from a temporary exploration reserve covering almost the entire Darling Range. Within a year a large tonnage of 32 per cent bauxite had been

Alcoa alumina refinery at Kwinana, about 1970.

located and a new company, Western Aluminum N.L., in which Broken Hill South Ltd and North Broken Hill Ltd joined WMC, was formed to develop an integrated aluminum industry. Sites were chosen for a mine and refinery in Western Australia and in Victoria for a smelter and a power station based on a nearby brown coal deposit.

All that was lacking was finance and technical know-how. Sir Lindesay Clark, AC, KBE, CMG, MC, Chairman of WMC and Broken Hill South Ltd and a Director of North Broken Hill Ltd, invited the Aluminum Company of America to take a 51 per cent interest and to supply the capital and technical expertise needed. Alcoa of Australia Ltd was formed in June 1961 and construction began soon after.

During 1963 bauxite mining at Jarrahdale, alumina refining at Kwinana, and aluminium smelting at Point Henry in Victoria began at annual rates of 600,000 tons, 200,000 tons and 20,000 tons respectively. The subsequent bringing into production of bauxite mines at Del Park, Huntly and Willowdale, the expansion of the Kwinana alumina refinery, and the building of new refineries at Pinjarra (1972) and Wagerup (1984) established Alcoa as the world's largest alumina producer, with a capacity in 1990 of 5,500,000 tonnes per annum. Expansion is under way at Wagerup, to increase the capacity to 6,130,000 tonnes per annum by 1993.

Worsley Alumina

Worsley Alumina Pty Ltd, owned in 1990 by Reynolds Australia Alumina Ltd (50 per cent), Shell Company of Australia Ltd (37.5 per cent) and Kobe Alumina Associates (Aust) Pty Ltd (12.5 per cent), explored for bauxite on the eastern side of the Darling Range and set up the Worsley Joint Venture in 1980. Construction of the refinery at Worsley and the mine at Saddleback began in September 1980. Alumina production started in April 1984, and by 1990 was at the rate of 1,200,000 tonnes per annum, all of it exported. The plant capacity is being increased progressively to 2,000,000 tonnes per annum and the joint venturers were examining the feasibility of an aluminium smelter in the Collie region.

Economic Contribution

In 1989 Alcoa's net contribution to Australia's balance of payments was $4,500,000 per day. In addition to export income, Alcoa and Worsley annually contribute about $825,000,000 to the Western Australian economy. Direct employment in 1990 was 4,800, and according to Government statistics indirect employment was at least four times the direct employment.

One remarkable achievement of the bauxite and alumina industry in Western Australia is its rehabilitation of mined areas. Alcoa was honoured in 1990 by the United Nations Environment Programme for its work in rehabilitating mined bauxite areas. Alcoa was the first Australian company, and the only mining company in the world, to be recognised in this manner. Similar work is done by Worsley Alumina.

The alumina refinery at Worsley is an important export earner.

Gold

The Boddington gold deposit, discovered in 1975, formed by the mobilisation of gold into the lateritic profile from a possibly uneconomic source within the bed rock, has a proven reserve of 45,000,000 tonnes at 1.8 grams per tonne.

Mining operations began in May 1987 at a rate of 3,000,000 tonnes per annum and the treatment plant was commissioned in July. It is managed by Worsley Alumina Pty Ltd on behalf of the joint venture partners. Production was increased to 6,000,000 tons per annum, yielding 433,100 ounces of gold, in 1989-90 and making it the largest individual gold mine in Australia.

Alcoa's Hedges gold mine works a similar low-grade gold reserve in the lateritic cap, alongside Boddington. Gold production commenced in October 1988 with an annual rate of production of 160,000 ounces.

Education

Education and training are vital for a viable mining industry. At first, Collie employers provided on-the-job training but, as the coal industry grew and people became more mobile, more formal training became desirable. Evening technical training classes were introduced at Collie in 1950 by what later became TAFE. The requirement for certain supervisory personnel in coal mining to have statutory qualifications and the expected expansion of the coal mining industry prompted the companies to press for a branch of the Western Australian School of Mines to be established in Collie. This was achieved in 1981 through the establishment of the Collie School of Mines, which become the Collie Federated School of Mines with the incorporation of TAFE.

Future Outlook

Since the late 1950s, the south-west of Western Australia has become one of the great centres of the mineral industry in Australia. It is a world leader in bauxite, alumina, and mineral sands production. Its contribution to Australia and Western Australia is enormous and this will continue from presently known resources well into the 21st century.

I am indebted to Gilbert Ralph, MBE, for the preparation of this paper, which was presented at Bunbury on 5 December 1990.

South Australia

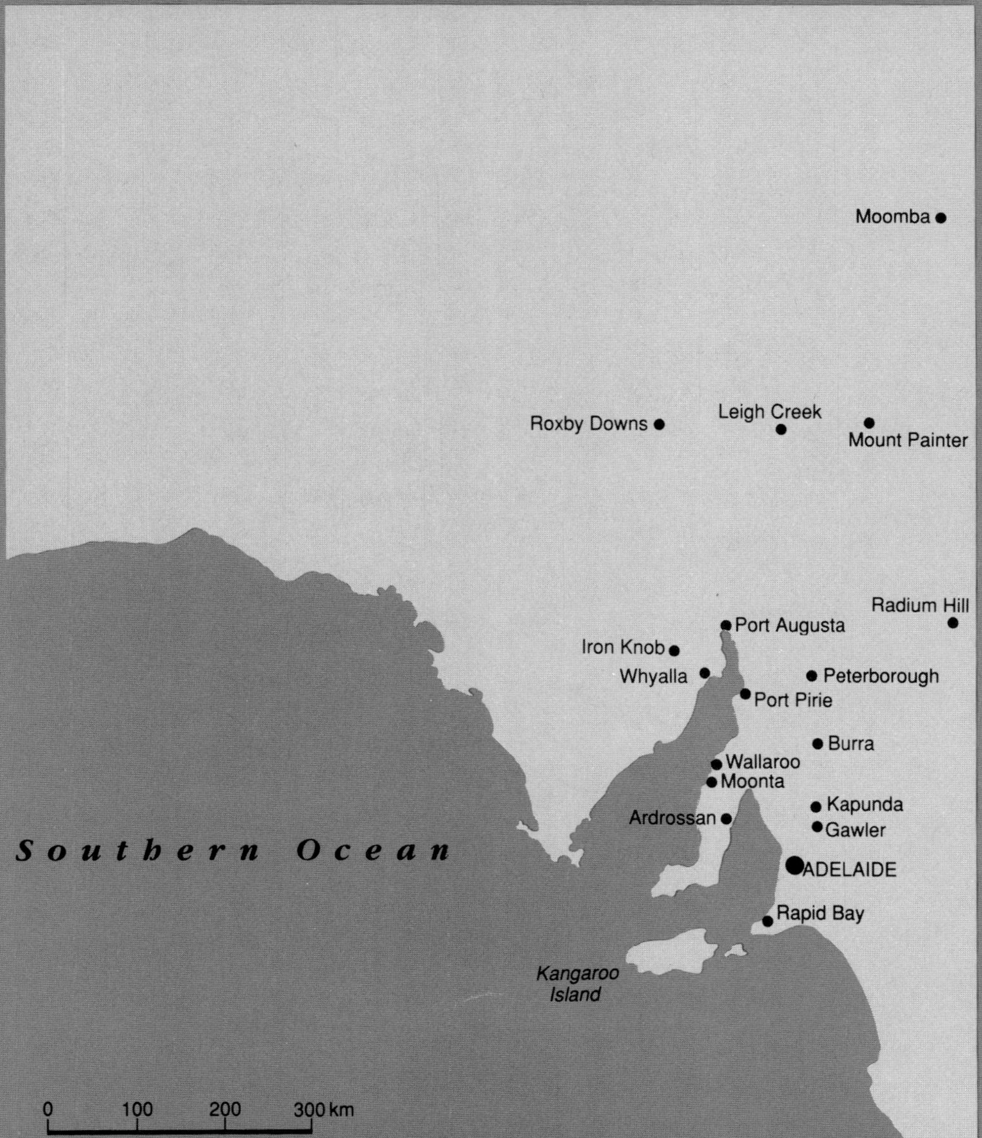

Education for the Mineral Industry

Mining in Australia began soon after the first settlement of New South Wales at Sydney Cove in 1788 with the quarrying of stone for building and construction. Within a decade coal was found near Newcastle and, later, on the South Coast.

Metal mining began with the discovery of silver-lead near Glen Osmond in South Australia in 1841. This, together with finds of copper at Kapunda in 1844, then Burra, and later in the Moonta-Wallaroo area, attracted miners and those experienced in smelting from Cornwall, Germany and central Europe.

Australia's great surge in mining came with the gold rushes of the 1850s. The initial finds of alluvial gold were soon replaced by the more difficult to mine quartz reefs. This again attracted qualified men from overseas.

Many were educated in Germany, where Mining Academies at Freiberg and Clausthal had been operating since 1765 and 1775 respectively. Early mineral industry professionals in English-speaking countries were trained at these Academies, including some of the Americans who came to Australia. In Britain a School of Mining was started in London in 1851. In 1863 it became the Royal School of Mines, under which name it still operates. Mining schools were also set up in Cornwall and Devon in the late 1850s, the most famous being the Camborne School of Mines. Graduates from these schools joined the professionals in the growing mining industry in Australia. In America, Columbia University opened its School of Mines, now called the Henry Krumb School, in 1864, followed by the Colorado School of Mines in 1874.

A. Mica Smith, Professor of Chemistry and Metallurgy at Ballarat School of Mines 1881-1912, Professor of Metallurgy 1912-1922.

Australian Schools of Mines

In Victoria, a Royal Commission in 1869 heard evidence of a lack of suitably trained technical people in the mining industry. Members of Parliament from the goldfields electorates were '... untiring in their advocacy of the cause of technical instruction'.

As Haughton H. Dunkin and Howard K. Worner, CBE, record in their detailed paper on *Mining and Metallurgical Education in Victoria*, published in The Institute's Proceedings in 1950, the setting up of the first School of Mines in Australia at Ballarat 'was initiated on 6 October 1869, at a meeting of the Mining Board, elected by holders of Mining Rights'. The inaugural address was given by the School's president, Sir Redmond Barry, Chief Justice of Victoria, on 26 October 1870 and classes commenced in January 1871 with four students.

Over the next two decades ten establishments with the name School of Mines were set up in Victoria, but more than half of them were really Technical Schools. Those which were near major mining centres and

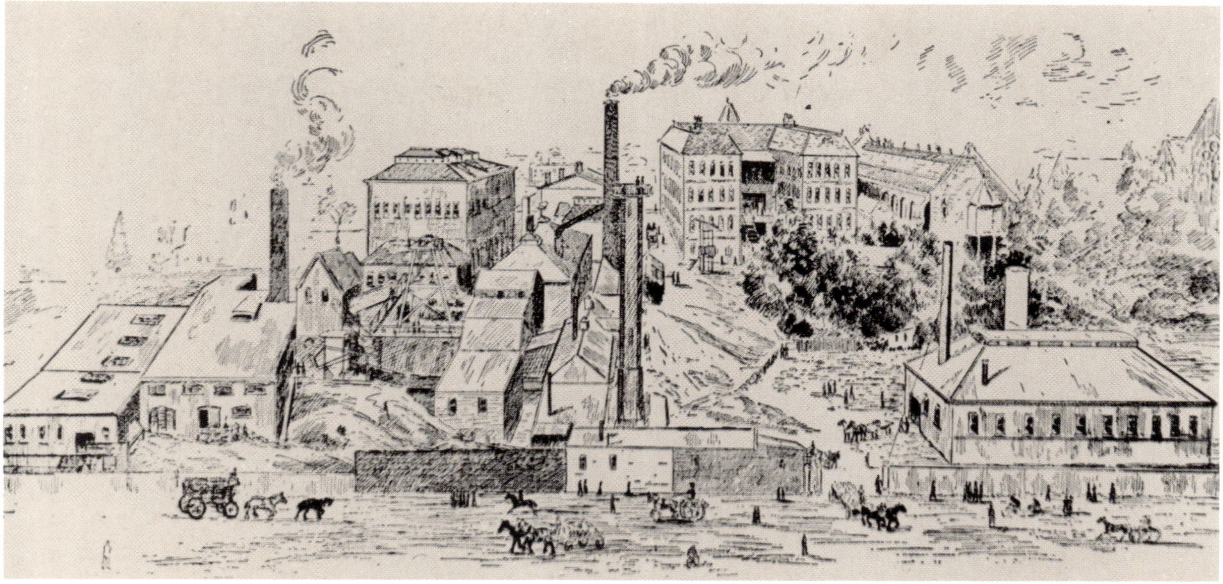

The School of Mines, Ballarat, about 1900. Rear view, showing the Battery, Assay Laboratory and Gauge Tower.

offered instruction in mining-related subjects were opened at Bendigo in 1873, Castlemaine in 1887, Maryborough in 1888, Bairnsdale in 1889, and Stawell in 1890.

The highly respected A. Mica Smith, who had been the first lecturer at Bendigo in 1873, was appointed Professor of Chemistry at Ballarat in 1881 and for over 40 years had a significant influence on students who entered the mining industry.

The Working Men's College in Victoria , which was established in 1887 and in time became the Royal Melbourne Institute of Technology, came into existence largely as a result of the vision, initiative and philanthropy of the Hon. Francis Ormond. He was succeeded as President by William C. Kernot. F. A. Campbell, appointed Principal in 1887, introduced instruction in assaying which became the basis of the School of Mines within the College. One of the first part-time instructors was B. N. F. Ulrich, a graduate of Clausthal, with wide experience in Europe, who at the time was a Field Officer with the Victorian Geological Survey.

A.A. Robertson, who began instruction and practical classes in Physical Metallurgy at Melbourne Technical College in 1935.

By the 1880s South Australia had extensive copper mining and smelting developments and, with the discovery of silver, lead and zinc at Broken Hill in 1883, the need for teaching institutions to provide trained personnel for the mines and smelters became more and more urgent. The South Australian School of Mines and Industries was started in Adelaide in 1889, the only real School of Mines in a capital city. The fine building on North Terrace, near the University, which for many years was the centre of mining education in South Australia, was built largely as a result of a gift from George Brookman, one of the promoters of Kalgoorlie's richest mine, Great Boulder. During the next few years five other institutions with the name School of Mines were set up in South Australia's major mining centres. One such school was established at Kapunda in 1891 and, like most of the others, came under the control of the South Australian School of Mines until 1915, when they were

School of Mines, North Terrace, Adelaide. Started in 1889, it was the only real School of Mines in a captial city.

Students from the WA School of Mines, plane table survey field work, 1951.

renamed Technical Schools and placed under the direct control of the Education Department.

The Western Australian School of Mines had its origins in 1896, three years after the discovery of gold in Kalgoorlie. Goldfields historian Norma King records that:

> In May of 1896, an editorial appears in *The Kalgoorlie Miner* suggesting that the Government should be asked for a grant to establish a School of Mines. Little came of the suggestion at the time but by 1899 some qualified mining men in Kalgoorlie and Boulder were conducting private classes in such subjects as mechanical and mathematical drawing, assaying, practical chemistry and metallurgy.

When Coolgardie and Kalgoorlie both made representations to the Government, it chose Coolgardie, and the Western Australian School of Mines was set up there in November 1902, the first Director being F. B. Allen, a 35-year-old New Zealander who had previously been Director of the Thames School of Mines in New Zealand. The School was transferred to Kalgoorlie late in 1903.

In Queensland, A. J. Wilson records that John Nash's discovery at Gympie in 1867 not only saved Queensland from financial disaster, but also gave rise to the first organised teaching by a local mining institute on how to 'prospect' and 'test on the spot'. But, he notes, 'Gympie's importance as an educational centre was short lived'. Other Schools of Mines were later set up at Charters Towers and Mount Morgan and for some years were important to mining education in Queensland. The Charters Towers School of Mines became world-renowned as a centre of research into both chlorination and cyanidation for the recovery of gold.

School of Mines, Charters Towers.

Sydney Technical College, founded early in the 1880s, introduced Diploma courses in mining and metallurgy in 1892. For many years it had a strong link with the Broken Hill College. At Broken Hill, the

Technical College opened early in the 1900s and for many years rivalled the Western Australian School of Mines in both reputation and number of diplomates.

Tasmania boasted Schools of Mines at Mount Lyell and Zeehan before 1900 and others followed at other mining centres later, but were short-lived.

In New Zealand, at that time an important gold producer, the School of Mines at Otago opened in 1879 under the direction of Professor B. N. F. Urlich. Others sprang up in mining centres such as Thames and Waihi. In 1986 the Otago School of Mines was closed and teaching in mineral industry disciplines transferred to Auckland University.

Almost all the Schools of Mines offered, in addition to the Diploma courses in mining, metallurgy and geology, instruction in sub-professional level with Certificate Courses in subjects such as surveying, assaying, battery operation, drafting and engineering. In many instances the courses were tailored to suit the needs of local industry, be it gold or copper, dredging or smelting.

The Broken Hill Technical College.

Universities

The few universities in Australia in the 1800s were slow to introduce degrees in mining or metallurgy. The Schools of Mines generally catered

No. 4158

SOUTH AUSTRALIAN SCHOOL OF MINES AND INDUSTRIES.

This is to Certify that Essington Lewis during the three Terms of the Year 1903, attended a Course of Instruction in Mining 1 and passed the Examination in that Subject in the Second Class at the Ordinary Examination, 1903.

Langdon Bonython, President of Council.

H. Haslam B.Sc. Registrar.

Dated this 24th day of December 1903

1903 certificate stating that Essington Lewis, destined to become Chairman of BHP, had studied Mining 1.

Western Australian School of Mines Laboratory, Kalgoorlie, 1930s.

for the practitioners in such fields as mining, assaying or chemistry, awarding diplomas of an acceptable standard for the mining companies, which were mostly dominated by graduates from England, Europe and, later, the United States.

The University of Melbourne, which opened in 1855, offered a mining option in 1874 as part of the Civil Engineering course. Lectures in metallurgy began in the early 1890s and formal degrees in mining were available from the turn of the century. The University of Sydney, which began teaching in 1852, started to offer engineering degrees in 1885, with mining and metallurgy options from 1892. By the beginning of this century these courses were well developed.

In South Australia, the University of Adelaide was founded in 1874, following a gift of £20,000 from Captain Walter Watson Hughes and later £100,000 from Sir Thomas Elder, GCMG, both of whom had made fortunes from the Wallaroo-Moonta copper mines. The University shared teaching facilities with the South Australian School of Mines for many years, until mining, metallurgy and applied geology teaching were concentrated in the Gartrell School of Mines within the South Australian Institute of Technology, now the University of South Australia.

The University of Tasmania, incorporated in 1889, introduced provisional courses in mining and metallurgy in that year, but so few students

David Phillips, appointed Foundation Professor of Mining Engineering at New South Wales University of Technology in 1949.

applied that the courses were discontinued in 1905 and only geology was available. Soon after its formation in 1911 the University of Western Australia introduced geology within its Science Faculty but it did not teach mining or metallurgy, these being more appropriate to the School of Mines in Kalgoorlie. A succession of notable Professors of Geology included E. De Clarke and Rex Prider, who forged close links with industry.

The University of Queensland, founded in 1909, did not begin mining courses until the late 1940s, with Professor Frank White appointed as the first occupant of the Chair of Mining Engineering in 1949. This followed the initiative of Julius Kruttschnitt of Mount Isa Mines, J. Malcolm Newman, CBE, of Mount Morgan, and Ian Morley, ISO, State Mining Engineer. One of the first graduates was Charles Copeman, Queensland Rhodes Scholar, who was President of The Institute in 1988.

Sir Rupert Myers, KBE, (centre) with visitors from the Sydney Branch of The Australian Institute of Mining & Metallurgy, at the New South Wales University of Technology, 1955. Myers became first Professor of Metallurgy there in 1951. The University later became the University of New South Wales.

Transition in Education

In the early 1900s the Schools of Mines began to change. In addition to mining and metallurgy, courses in other branches of engineering and chemistry became available. This was in line with the growing importance of secondary industries, a development greatly accelerated by the First World War, during which the steel industry and other manufacturing industries were established. The decline in gold mining in Victoria also had a bearing on the change. Many of the former Schools of Mines, which had dropped the teaching of mining and metallurgy at professional levels by 1902, changed their names to Technical Schools.

Official opening, School of Mines, Otago University, New Zealand, 1909.

The University of Melbourne Engineering and Geology Schools. The University offered its first Mining option in 1874 as part of the Civil Engineering Course.

Physical Metallurgy had developed over the first two decades of the 1900s, mainly in Europe (including Britain) and the United States.

A Chair of Metallurgy was created at the University of Melbourne in 1923, with Dr J. Neill Greenwood, a graduate of Manchester, as first Professor. He was the only Professor of Metallurgy in Australia until 1939, although W. L. Allen began instruction and practical classes in Physical Metallurgy at Footscray as early as 1923, followed by A. A. Robertson at the Melbourne Technical College in 1935.

Not until the late 1930s was attention given in New South Wales technical colleges to Physical and the Secondary Metallurgy. This was surprising, seeing that the steel making plants were in New South Wales. Immediately before and during Second World War a fine technical college was built in Newcastle, including a Physical Metallurgy Department. Wollongong and Sydney Technical Colleges also stepped up their Physical and Secondary Metallurgy courses during the war years.

David Reeve, a mining student at the University of New South Wales, receives the Atlas Copco Travelling Scholarship for Mining Undergraduates from 1990 Australasian Institute of Mining & Metallurgy President Sir Arvi Parbo.

Sydney Technical College became the University of Technology in 1949 and later the University of New South Wales. Professor Sir Rupert Myers, KBE, was the first Professor of Metallurgy in 1951 and Professor David Phillips began Mining Engineering the same year. He was succeeded by Professor John Morgan from Adelaide. Morgan advanced the School of Mining and, in addition to his academic achievement, was a great supporter of The Institute from the time he won the Student Essay Competition in 1935. He was President of the Institute in 1981 and recipient of the Institute Medal in 1987.

Professor Greenwood's students occupied Chairs in several universities: Professor Frank White in Queensland, Sir Rupert Myers, KBE, in New South Wales, and Howard Worner, CBE, at Melbourne after Greenwood retired to take up a research professorship. Professor Hill Worner, also a student of Greenwood and brother of Howard, succeeded Howard to the Chair in Melbourne. Between them, the Worners headed Metallurgy at Melbourne for 30 years.

Geology was first taught by Frederick McCoy at Melbourne in 1854 and, on his death in 1899, Professor J. W. Gregory was appointed to the Chair of Geology and Mineralogy. He resigned in 1904 and Professor E. W. Skeats held the post until 1941. He influenced the teaching of geology throughout the nation and was President of The Institute in 1925. Geology was first taught at various institutions as part of Science. Australian mining fields were almost devoid of practicing geologists until the 1930s. Careers for geologists were mostly with the various Geological Surveys.

In Adelaide, H. W. 'Spog' Gartrell led the teaching of mining and metallurgy at the University of Adelaide and the South Australian School of Mines from 1910 until his death in 1945. He was the first Professor of Mining and Metallurgy, in 1939, and was highly respected by his students and his peers. Professor Gartrell was a Councillor of The Institute from 1939 to 1945 and President in 1941. Among his many students were Sir George Fisher, CMG, William M. Morgan, CMG, Sir Frank Espie, OBE, Sir Russel Madigan, OBE, (Professor) Cedric Gregory, and (Professor) John Morgan. He was succeeded by Professor E. C. R. (Clinton) Spooner from Tasmania.

Notable in geology at Adelaide were Professor Sir Douglas Mawson of Antarctic fame, and Dr Cecil T. Madigan, desert and Antarctic explorer. After a career in industry, Professor Eric Rudd, AC, was appointed to the industry-sponsored Chair of Economic Geology in 1949.

Many of the professors in mineral industry disciplines at Australian educational institutions have enjoyed and continue to enjoy world-wide reputations.

Dorothy Hill, who graduated in Geology at Queensland University in 1929, culminated a distinguished career in research and training as Research Professor there.

Professor Albyn Lynch, AM, Queensland University, President of The Institute in 1989, was instrumental in establishing the AusIMM Education Endowment Fund.

Geology Department, University of Queensland.

Geology students from Queensland University on a field excursion to The Gap Quarry, 1911.

Over the years, the mining industry has supported the Schools of Mines and Universities by gifts and various endowments. Mount Isa Mines Ltd was mainly responsible for the establishment of the Julius Kruttschnitt Research Centre, in Queensland, which is now world famous.

In addition to teaching, research in mineral industry technology has been pursued successfully at most Australian Universities and Schools of Mines. Many Australian technological discoveries and innovations have found world-wide application.

Professor Howard Worner, CBE, Chairman of The Institute's Education Committee for many years.

Education and The Institute

Great contributions have been made to The Institute by many of the educators. Professors E. W. Skeats, J. N. Greenwood, H. W. Gartrell, E. A. Rudd, AC, J. P. Morgan and A. J. Lynch, AM, were all past Presidents and Professor H. K. Worner, CBE, was Chairman of The Institute's Education Committee for many years. Various industry leaders and former Presidents, such as Sir George Fisher, CMG, Sir James Foots, Sir Henry Somerset, CBE, and Sir Laurence Brodie-Hall, CMG, have been or are Chancellors or Chairmen of their respective Universities or Schools. Among Institute Medallists to receive awards for services to education were Geoffrey O'Malley in 1957, Neill Greenwood in 1961, Eric Rudd, AC, in 1968, Howard Worner, CBE, in 1977 and John Morgan in 1987.

Graduates in the Industry

Professor Howard Worner, CBE, estimates that, from the start of the 20th century until the First World War, the annual average number of diplomates in mining and metallurgy from Australian Schools of Mines was about 15 in mining and 25 in metallurgy. The South Australian School of Mines had the largest number in mining, followed by Ballarat, and then Western Australia. In metallurgy, most came from Melbourne, followed by South Australia and Ballarat. Universities turned out about the same total number of graduates, with most mining engineers coming from Adelaide and Sydney and metallurgists from Melbourne.

In more recent times, from 1981 to 1985, the average number of graduates from Australian institutions in mining engineering annually was 62, with 40 in metallurgy and chemical engineering.

Professor J.N. Greenwood, first Professor of Metallurgy at The University of Melbourne.

It is extremely difficult to obtain reliable figures for numbers of professionals employed in the industry. The Australian Mining Industry Council/The Australasian Institute of Mining and Metallurgy survey in 1985 indicated that there were 1,027 mining engineers, 798 metallurgists, 364 chemical engineers, and 1,321 geoscientists.

In mining engineering 147 graduated in 1985, of whom 69 were from the University of New South Wales. In metallurgy and chemical engineering there were 231 graduates, and in earth sciences 475 graduates.

About half the mining engineering graduates go to mining companies, according to the best estimates. Many are also attracted to Government departments and instrumentalities, and a few for various reasons do not follow their professions. In metallurgy and chemical engineering the picture is even more obscure, but only about 20 to 30 per cent go into the mineral industry.

The AusIMM Education Endowment Fund

The AusIMM's Education Endowment Fund was established in April 1989. At 19 July 1990, a progressive total of $3,546,990 had been committed towards the target of $5,000,000, including $52,000 by five Branches of The Institute and over $30,000 by over 130 individual members. The first scholarships from the Fund were awarded to high calibre students in disciplines relevant to the mineral industry.

The fund needs and deserves more support. Nothing is more important to our future than ensuring first class successors to today's professionals.

Sir Arvi Parbo (centre) with students at Sydney AusIMM Branch meeting, 1990.

The present becomes the past sooner than we like. Leadership of The Institute and the industry will soon be in the hands of today's students. Their success, and the success of the industry, depends on our efforts to encourage the best young people to train for the professions.

Professor H.W. Gartrell, after whom the Gartrell School of Mines at the University of South Australia is named. He was President of The Institute in 1941.

We all know that education is far from complete on graduation. A degree or diploma is merely a passport into the world of opportunity and experience. Professor Gartrell in his 1941 Presidential Address said:

> Engineering might, with a considerable amount of truth, be briefly defined as doing a prescribed job at a minimum cost, so, early in his course, the student should have something of the sort impressed on him, together with the fact that safety is a constant requirement and that cost includes amortisation. It is not our business to turn out even budding executives, but if design is taught with maintenance in mind, if flow sheets are taught with due regard to overhead charges, if instructor after instructor stresses the financial side of his work, the young graduate will go out prepared to continue this side of his education.
>
> This brings us to the true test of curriculum and of teacher alike; it lies in the answers to two questions, 'How well prepared is the young graduate to learn for himself?' and 'How keen is he on continuing to learn'?'

Professor Gartrell's words remain true today, half a century later, except that the young graduate might also need to be prepared to learn for herself as well as himself.

I acknowledge gratefully the invaluable contributions of Professor Howard Worner, CBE, John Dew and Gilbert Ralph, MBE, to the preparation of this paper, which was presented in Adelaide on 14 August 1990.

South Australia's Wealth in Copper

South Australia has an important place in Australia's mineral history. It was the cradle of metal mining in Australia, it saw mineral production ensure its economic success, it ranked high in the world as a copper producer early in its history, it produced the nation's first uranium, and it is the site of a remarkable recent mineral discovery. It is also the birthplace of The Institute, which was initiated in Broken Hill in 1892 and formally established in Adelaide in 1893.

The discovery of galena in February 1841 at Glen Osmond in the foothills of the Mount Lofty Ranges, near Adelaide, soon led to the development of Australia's first metal mine, and provided hope of extra income for South Australia.

Historian Derek Whitelock in his book *Adelaide* describes how the nation's first paid mineralogist, Professor Menge, emerged from the bush a number of times to declare through public lectures and at the Mechanics Institute that 'the province was rich in minerals'. Small deposits of silver, lead, copper and gold were reported in the Adelaide Hills.

Silver-lead mine, Glen Osmond. Discovery of galena in the Adelaide Hills in 1841 led to Australia's first metal mine, and provided hope of extra income for the recently-settled South Australia, then struggling to establish its economy.

In 1842 pastoralist Francis Dutton found outcrops of what he described as 'a beautiful green moss' on his property at Kapunda, about the same time as a similar discovery on the next door property of Captain Bagot. Whitelock records that, two years later, 'after prayers and a learned address by Professor Menge, the Kapunda copper mine was opened'.

The Kapunda workings were shallow but the flow of water was so great that a steam-driven pump (one of the first in the Colony) was installed in 1848 to dewater the workings. The first smelter was fired in 1849. The field operated at a rate of about 2,000 tons of hand-picked ore per annum until 1886, after which it was worked by tributers until 1912. Although a small mine the ore was rich and the revenue important to the struggling Colony. Total production was about 15,000 tons of copper.

In 1845 two shepherds independently discovered copper near what is now Burra. This was beyond the properly surveyed districts and land could only be acquired as 20,000 acre blocks, and at a price of one pound per acre, 'in gold coin'. Two rival syndicates acquired adjoining blocks. One had no minerals of value, the other the Burra copper deposits. The Burra mines brought wealth to many, including Henry Ayers, the Mine Secretary, who later entered politics, became Premier, was knighted, and whose stately home, Ayers House, on North Terrace remains a city landmark. Sir Henry in 1893 became foundation President of the Australasian Institute of Mining Engineers, the forerunner of The Australasian Institute of Mining and Metallurgy.

Remains of silver-lead smelter at Glen Osmond, 1950.

By 1850, about 5,000 people lived in and around Burra, many under the auspices of the South Australian Mining Association, which owned

Kapunda copper mine. Copper was discovered here in 1842, the first smelter fired in 1849. The mine operated at a rate of about 2,000 tons of ore per annum until 1886, after which it was worked by tributers until 1912. Total output was about 15,000 tonnes of copper.

Cornish miners at Moonta, 1895.

the mine and adopted the typically Cornish practice of providing welfare and community facilities, such as housing, water supply, churches, a hospital, and even a cemetery. It was indeed a benevolent and prosperous society. The great Monster Mine alone produced over £750,000 worth of copper before Victoria produced an ounce of gold. Dividends were generous and £10 shares sold for £200.

Mineral production saved South Australia from bankruptcy, and made it the most prosperous of the Australian colonies by the middle of the 1800s.

Although the gold discoveries in the eastern colonies eclipsed the role of copper in South Australia, it continued to provide the colony with prosperity. During the summer of 1859-60, retired sea Captain Walter Watson Hughes engaged four Cornish miners from Burra to sink a shaft near a spot on his sheep property, near Wallaroo, where his shepherd, James Boor, had found carbonate of copper some years earlier. They chose the site by the traditional Cornish method of 'swinging a pick around the head and sinking the shaft where it landed'. As if to confirm the wisdom of this practice, the shaft met a rich copper lode at a shallow depth.

Needing capital, Hughes invited Sir Thomas Elder, GCMG, and Robert Barr-Smith, of Elder Smith fame, to join him. This led to development of the copper deposits in the Wallaroo, Kadina, and Moonta area. Large numbers of Cornish miners from Burra and the then-declining Kapunda mines came to the area and it soon became known as 'Australia's Little

Cornwall', graphically described in a book of that name by cartoonist Oswald Pryor.

Notable among its people was Captain Henry Hancock, affectionately known and respected as 'Cap'n 'ancock'. Sir Ben Dickinson wrote of Hancock that he was:

> an autocrat in complete control of day to day decision making for more than thirty years. He knew every aspect of hard rock mining. He was an innovative engineer who developed the hydraulic jig which found its place in mineral processing all over the world. He was thoroughly conversant with smelting. He was a man who bridged the gap between the owners and the workers. He knew something about dividends, profits and losses. The owners gave Henry his head. They loved him; so did the miners ... He recognised that the miners had a privilege to strike for higher wages but he made it clear that there would be much suffering amongst those who had saved no money when he could not grant their demands. The miners remembered the innumerable kindnesses he had shown them when sickness, disabilities and trouble had afflicted their families and they usually reported back to work.

The output of these three centres, together with the 300-or-so smaller mines throughout the Mount Lofty and Flinders Ranges, established South Australia as a major copper producer, which at its peak supplied

Ore pickers, Moonta, 1913. Such work was usually done by boys.

Morphett's shaft at the Burra copper mine, 1906. The cut-out on the headframe is Johnny Green, the miner's mascot.

Weighing refined copper into store ready for shipping, Wallaroo smelting works, about 1915.

some 10 per cent of the world's needs. The Colony was able to take advantage of the rapid increase in the world demand for copper wire during the second half of the 19th century, which resulted from the introduction of telegraphs, telephones and electricity.

Numerous relics of South Australia's early copper mining days have been preserved or restored, notably Morphett's engine house at Burra and the Hughes pump house and chimney at Moonta. Preservation of these and other relics is a vital part of our mineral heritage.

Prosperity also accrued to South Australia from the secondary industries set up to serve the mines. Typical of these, the firm of Hawke and Co, at Kapunda, became a very prominent engineering foundry and weighbridge manufacturer. In nearby Gawler, James Martin & Co. had a large engineering works, initially to satisfy the mining industry, and later to build locomotives and other heavy engineering equipment, mainly for South Australia.

Gold

Copper discoveries in South Australia were followed by modest discoveries of gold. The *Royal South Australian Almanac* for 1848 mentions that 'the Victoria Goldmining Company possessed two Sections comprising 147 acres. Soon after operations commenced in January 1846, a vein of auriferous gossan was discovered impregnated with gold of almost perfect purity'. The company produced 24 ounces of gold, believed to be the first in South Australia.

The exodus of 20,000 people from South Australia to the Victorian gold rushes from 1851 had a greater impact than merely the loss of the

Copper smelters, Wallaroo, 1870s. They treated all the ore from Moonta and Wallaroo.

best miners. Much South Australian money went, too, and there was a desperate shortage of coin. This was remedied in 1852 when the Bullion Act made gold bullion legal tender and, as the South Australian Assay Office offered a higher price than Victoria, a great deal of gold flowed from Victoria to Adelaide. Many South Australians earned money out of gold indirectly, as suppliers of goods and services such as food, machinery and champagne to the Victorian fields. Little gold was ever found in South Australia, the largest field, Waukaringa, near Yunta, producing less than 100,000 ounces.

South Australia had one other connection with the Victorian gold rushes. In 1857-58 some 16,500 Chinese arrived at the port of Robe, bound for the Victorian goldfields at Ballarat, hundreds of kilometres away.

Victoria levied a poll tax of £10 for each Chinese immigrant and forbade more than one Chinese passenger for each ton of the vessel's deadweight. Landing at Robe avoided the tax. The locals were quick to cash in on the situation, charging five shillings per head to bring the new arrivals ashore and £50 per party to escort them to Ballarat. Thomas Smeaton in his book *The Great Chinese Invasion*, published in 1865, claimed that many of the guides vanished after only a day or two, leaving the Chinese on their own in the wilderness.

The Talisker silver-lead mine, 1870.

Talisker Mine

The Talisker silver-lead mine near Cape Jervis was discovered by brothers Don and John McLeod, who were shepherds in the area. They found the outcrop in July 1862 and mining began soon after under the leadership of Captain Alfred Jenkin, who sailed from Glenelg to the Cape with a party of 20, including some experienced Cornish miners. Operations reached a peak about 1867. Total production amounted to some 900 tons of lead bullion which was shipped to England.

Transport has always been a problem for outback miners. Transferring supplies from motor lorry to camels near Blue Creek Mine, 1926. The lorry carted ore and concentrates to Copley for railing to the radium extraction plant at Dry Creek.

A boiler on its way from Kapunda to the Talisker silver-lead mine.

Partially denuded of trees to provide timber for mine support, fuel, and building materials, the area has been rehabilitated by natural regrowth and in 1987 was declared a Conservation Park.

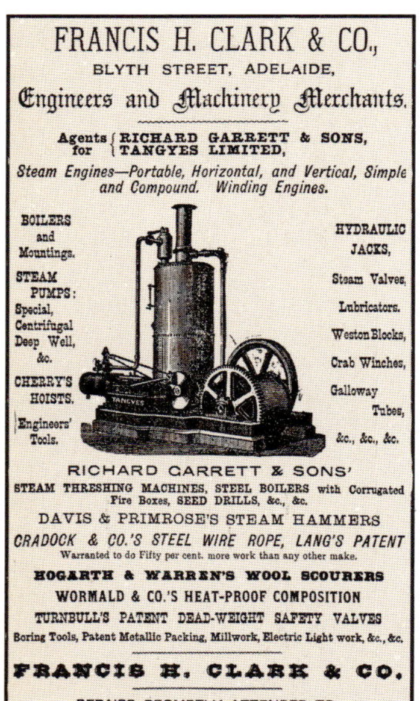

South Australian mines provided a steady trade for machinery merchants and manufacturers.

Leigh Creek Coalfield

Coal-bearing shale was discovered near Copley in 1888 during the sinking of a railway dam. Drilling and shaft sinking confirmed the existence of several coal seams. In 1892 the Leigh Creek Mining Co. was formed and some 200 tons of coal raised, chiefly for experimental purposes. The field was soon abandoned.

When coal shortages during the Second World War prompted a further examination of the field, the large shallow Northern Basin was discovered and developed in 1943 by the Engineering and Water Supply Department. In 1948 the Electricity Trust of South Australia took over large-scale development of the field to supply the Osborne Power Station near Adelaide and later the Port Augusta Power Stations. Leigh Creek coal has provided a reliable electricity supply to South Australia for more than 40 years.

Uranium

South Australia produced Australia's first uranium. In the early 1900s, A. J. Smith, a prospector, discovered what he thought was a rich tin deposit, but which on closer examination proved to be uranium. It became the source of radium for glass pigmentation and in medical research by Madame Curie and others. The place was called Radium Hill and the uranium mine there operated on and off until about 1930, producing a few hundred milligrams of radium and a ton or two of uranium oxide by-product, before the rich Eldorado pitchblende deposits in Canada became the world's main source of radium.

The Second World War prompted renewed interest in the uranium deposits at Mount Painter and Radium Hill. Because of their strategic significance for weapons they were retained by the South Australian Government. After the war the potential for nuclear power generation led to the reopening of Radium Hill and a uranium treatment plant at Port Pirie, both as government enterprises. Production ceased in 1962.

Open-cut coal mine at Leigh Creek, in the north of South Australia. Mining began here in 1948 to help make South Australia less dependent on coal from other states. Coal is railed to powerhouses at Port Augusta.

Olympic Dam

One of the world's great mineral discoveries occurred in South Australia in 1975. It was made by Western Mining Corporation Ltd (WMC), which had been involved in copper exploration in South Australia, mainly in the Moonta-Wallaroo area, since 1959. In the early 1970s WMC's exploration staff postulated that deposits of copper in sedimentary rocks should occur to the west of the Flinders Ranges. After a year of reconnaissance, WMC decided to concentrate the effort north of Port Augusta in an area known as the Stuart Shelf, a vast expanse covered by sand dunes. Only a small copper deposit near Mount Gunson was known in the region.

Grinding ore with a roll crusher at Balhannah, in the Adelaide Hills, about 1870. The Balhannah copper and bismuth mine began operations in 1867.

Early copper mining in South Australia. Miners at Wallaroo copper mine, Kadina.

The possibility of favourable host rocks at depth had been shown by earlier magnetic and gravity surveys by the Bureau of Mineral Resources and this, together with a careful study of tectonics, led to identification of a number of targets. The first drill hole, near Olympic Dam, in July 1975, was at the 353 metre level when it encountered copper mineralisation which persisted for 38 metres and had a copper content of 1.5 per cent.

Eight more holes failed to find better intersections. The tenth hole, however, intersected mineralisation which assayed 2.1 per cent copper. Later drilling was to reveal a resource of more than 2,000 million tonnes containing copper, uranium, gold, and silver, all concealed by 350 metres of barren rock. From 1979 the deposit was developed as a Joint Venture between WMC and British Petroleum Ltd (BP). The decision to proceed into production was taken in December 1985 and production commenced in mid-1988.

As well as an underground mine, the operation consists of a copper concentrator, a hydrometallurgical plant for the recovery of uranium and some of the copper, a copper smelter, and a copper refinery with a combined precious metals plant, all incorporating the best-available technology. The combination of various technologies at Olympic Dam exists nowhere else in the world.

In 1989-90 the project produced 37,800 tonnes of refined copper, 1,000 tonnes of uranium oxide, and 18,000 ounces of gold. In 1990 the joint venturers were considering additions which would increase copper production to 66,000 tonnes per annum and uranium production to 1,400 tonnes per annum. Expansion beyond that depends mainly on world

Cartoonist Oswald Pryor made the Cousin Jacks and Jennys of Australia's Little Cornwall the subject of many a cartoon.

'What'd make the needle always point north, Cap'n?'
'Think I'm goin' tell you in five minutes what took me years to find out?'

markets for the products. At any likely rate of production, the Olympic Dam operation has a life of some hundreds of years.

A rock breaker at the Whenan shaft, Olympic Dam.

Santos

Santos Ltd, one of Australia's most successful petroleum exploration companies, was incorporated in March 1954. Its name was derived from South Australian and Northern Territory Oil Search.

Santos discovered natural gas in the Gidgealpa Wells and Moomba area during the 1960s. By the end of 1969, gas was being sold to the South Australian Gas Company in Adelaide. In 1976 a gas pipeline was completed to Sydney. During 1980-82 a liquids pipeline was built to the Stony Point processing and loading facility near Whyalla.

Australian Mineral Development Laboratories

The Australian Mineral Development Laboratories in Adelaide began as the Research and Development Branch of the South Australian Mines Department, first set up in the early 1950s to conduct research and pilot scale tests on Radium Hill uranium ore. The pilot plant at West Thebarton and the Chemical Laboratories at Frewville were contributed by the South Australian Government to form AMDEL, with further support from the Commonwealth Government and the Australian mining industry. The industry worked through the Australian Mineral Industries

Molten copper pours from an anode furnace into the casting wheel at Olympic Dam.

Research Association (AMIRA) with The Institute playing a crucial part in its formation in 1959. AMDEL later changed to a private enterprise organisation and became a recognised consulting service to the mineral industry in Australia and overseas.

Laying the Moomba-Sydney gas pipeline. Santos found natural gas in the Gidgealpa Wells and Moomba area during the 1960s. By 1969 gas was being sold to Adelaide and in 1976 a gas pipeline was completed to Sydney.

Australian Mineral Foundation

The Australian Mineral Foundation (AMF) was set up in May 1970 as a national earth sciences information and education centre. It was created by industries, institutions, individual benefactors, and government departments concerned with minerals and petroleum. Its extensive library and mineral industry related information system (AESIS) provides a unique service. Training programmes and specialist courses are conducted in a wide range of mineral and petroleum industry topics.

Future Outlook

The South Australian mineral industry recorded some remarkable successes in its early existence, including the honour of being the birthplace of Australia's metal mining industry. Its contribution to South Australia was vital to the State's economic and educational development.

After a quiet period in the first half of the 20th century, the industry again became a prominent contributor to South Australia. The State also became an important centre for mineral industry research and education.

South Australia has potential for further discoveries and mineral developments. For these to occur, encouragement and support by government and public opinion are essential. I believe that South Australians once again appreciate the vital role of the mineral industry in their State.

I am indebted to Gilbert Ralph, MBE, for the preparation of this paper, which was presented at Roxby Downs on 25 November 1990.

The Radium Hill mine which operated between 1954 and 1961 to supply uranium under government contract.

Mineral Processing at Port Pirie

The name Broken Hill Associated Smelters (BHAS) was for generations a constant reminder of Port Pirie's close association with the source of most of its lead feed: Broken Hill.

Towards the end of the 19th century the companies at Broken Hill had their own smelters. BHP, or 'The Proprietary' as it was then called, was the 'Big Mine' and had the 'Big Smelter' at Broken Hill, with 15 furnaces using ironstone flux from South Australia. The British Broken Hill Company had set up smelters at Port Pirie, and later BHP also began smelting at Port Pirie, as the ironstone flux came from across Spencer Gulf from Iron Knob, which then had the largest known iron ore deposits in Australia. A tramway carried the iron ore to what was to become Whyalla and barges took it across the Gulf to the smelters.

As the smelters grew, so did Port Pirie. Early in the 20th century it became South Australia's second largest town. Living conditions were difficult: unsealed roads, mostly draught horses for transport, mangroves, soil so salty that few plants would grow, and sulphurous fumes, which often enveloped the town, from the low chimney stacks.

Silver-lead bullion ready for shipment from Port Pirie, about 1915.

The story of how BHP came to withdraw from smelting at Port Pirie is of interest. BHP Directors met at Port Pirie in June 1911 and Guillaume Delprat, CBE, the General Manager, reasoned that as their ore reserves at Broken Hill were limited, the company should consider starting a steel-making business based on the vast iron ore resource the company held at Iron Knob and nearby. The Directors agreed to consider the proposal and Delprat had a comprehensive report prepared by a Philadelphia consulting engineer, David Baker. The project was agreed, work began in 1913 and by 1915 BHP was producing steel at Newcastle. To finance the development, they raised more than £1,000,000 through a share issue and sold a major interest in their Port Pirie Smelter to several of the other producers at Broken Hill who had more extensive ore reserves.

Australia was then the world's largest exporter of lead in concentrates and metal. Its three lead smelters (Port Pirie, Zeehan and Cockle Creek) had a capacity to smelt about half the concentrates produced.

The outbreak of the First World War proved awkward for three of the Broken Hill companies: (North Broken Hill Ltd, Broken Hill South Ltd, and The Zinc Corporation Ltd), who had been sending concentrates to Germany for smelting. Finding themselves suddenly unable to do this, they were keen to acquire an equity in the BHP Port Pirie Smelter.

As Australia became more involved in the war, the need for metals increased. In 1915 the three Broken Hill companies concluded their purchase of a major interest in the BHP smelter at Port Pirie, although it was another decade before BHP sold the then remaining interest.

The Gawler firm of James Martin & Co. became prominent suppliers of equipment to the minerals industry.

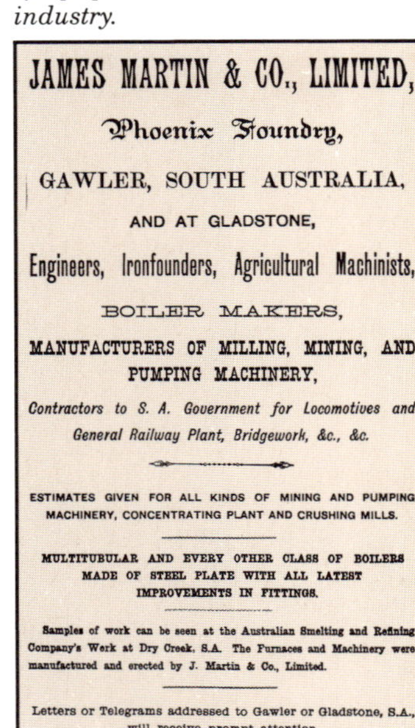

The Port Pirie lead smelters in 1903. Modern precipitators, fume control equipment and tall chimneys have eliminated this sort of pollution.

The Broken Hill Associated Smelters Pty Ltd was formed in May 1915 by W. S. Robinson. He and Sir Colin Fraser became Joint Managing Directors. In 1915 Robinson wrote to Fraser: 'We are not running this plant for the benefit of shareholders only . . . we are trustees in control of an important section of the requirements of the nations now struggling for their very existence. . . . I want lead and still more lead . . . damn the expense'. By the end of the First World War BHAS was the world's biggest producer of lead.

Sir Colin was a New Zealander who had graduated in geology and later gained considerable experience around the world before attracting the attention of W. L. Baillieu and W. S. Robinson. He became a powerful figure in what was known as the Collins House group and was Chairman of several of the companies. He was founding Chairman of Western Mining Corporation Ltd.

Sir Colin and W. S. Robinson were two of Australia's most distinguished mining leaders during the first half of the century. Both were later honoured by The Institute, Sir Colin as President in 1928 and Robinson as recipient of the Institute Medal in 1949.

The First World War also brought developments in zinc production. The only metallic zinc produced in Australia was from the BHP roasting furnaces and ovens at Port Pirie, but their capacity was only a fraction of that needed to treat the zinc concentrates available, most of which had been sent to the major zinc smelters in Belgium and Germany.

Amalgamated Zinc Ltd, which produced zinc concentrates at Broken Hill, using the de Bavay flotation technique, financed the setting up of the Electrolytic Zinc Company of Australasia Ltd (EZ) at Risdon, near Hobart, to produce metallic zinc by an electrolytic process developed in the United States in 1914 by the Anaconda Copper Mining Company. Sir Herbert Gepp, who was manager of Amalgamated Zinc, became EZ's first Managing Director.

EZ had a link with Port Pirie, because North Broken Hill, Broken Hill South and the Zinc Corporation were also involved in financing the new electrolytic zinc industry in Tasmania. By the end of the First World War the first zinc metal had been produced using the electrolytic process.

The BHP steelworks were established at Newcastle, because at that time it took three tons of coal to smelt one ton of iron ore, so it was more economical to take the iron ore to the coal. Later, however, backloading of coal enabled a blast furnace and eventually a ship building industry to be established at Whyalla.

Many Australian industries can trace their growth and development to the production of metals from Broken Hill and Port Pirie. Refined metals and sulphuric acid from Port Pirie provided Australian industry with an assured supply of raw materials, increased the value of exports and gave security and profit to the mines at Broken Hill.

From the Collins House group, in the two decades between the World Wars, came both the initiative and funds for other substantial industries,

Horizontal retort furnaces for zinc production at Port Pirie, about 1915. Port Pirie produced only small quantities of zinc before the First World War. Until then, zinc concentrates from Broken Hill were sent to smelters in Belgium and Germany.

Lead awaits shipment from Port Pirie, about 1917. This 30,000 tons of lead, the most ever stockpiled at one time, was needed for British Empire munitions production during the First World War. Lead mined at Broken Hill was critical to the Allied war effort and by the end of the war Port Pirie had the world's largest lead smelter.

such as paint manufacturing in Sydney (British Australian Lead Manufacturers Ltd), expansion of the copper smelters and refinery at Port Kembla (Electrolytic Refining & Smelting Co.), manufacture of brass and copper sheets, rods and cables (Metal Manufactures Ltd) and a major pulp and paper mill at Burnie in Tasmania (Associated Pulp and Paper Mills Ltd). It also joined with BHP, General Motors-Holden's Ltd and Orient Steam Navigation Co Ltd to form Commonwealth Aircraft Corporation Ltd to produce aircraft at Fishermen's Bend, Melbourne.

Restructuring of three companies involved with the BHAS saw various changes. Broken Hill South was absorbed into the WMC in 1980, with its BHAS and ER&S interests passing to the CRA Group, which owned the Zinc Corporation and New Broken Hill Consolidated. North Broken Hill took over EZ and APPM. In 1989 CRA and North Broken Hill rationalised their lead-zinc interests at Broken Hill, Port Pirie, Risdon, Elura, Cockle Creek and other related activities in Australia and overseas to form Pasminco Ltd.

A.J. Keast, CBE. He had a long association with Australia's silver-lead-zinc industry. He did much to improve the Port Pirie environment.

One of those who helped develop operations at Port Pirie in the early period soon after the Second War was A. J. Keast, CBE. A man with firm views, who had his detractors and his supporters, Keast had a long association with the lead-silver-zinc industry in technical and managerial roles, as well as making major contributions to other sectors of the Australian mineral industry. This was recognised by The Institute when he was awarded The Institute Medal in 1963. He was for many years a Councillor of The Institute and President in 1946. Based in Melbourne, he was a frequent visitor to Port Pirie. Those who worked at the plant in those days said his visits were always forewarned and the place was 'spick and span' when he arrived. He was very supportive of research and 'finding better ways of doing things'.

During the 1930s the Port Pirie smelters won a worldwide reputation for the continuous desilverising method in the refinery, developed under the leadership of Dr G. K. Williams.

Sintering of the Broken Hill concentrates before they were fed to the blast furnaces was, and remained, an integral part of the process. Port Pirie's two sintering machines were the largest of their kind in the world, and it was planned to build a third machine. Keast in his privately-published memoirs *Straws in the Wind* wrote that down-draught sintering, developed in the early 1900s by two Americans, Dwight and Lloyd, had never changed materially from the original concept.

The hard, fines-free, low-sulphur sinter produced by updraught sintering was an important factor in the zinc blast furnace known as the Imperial Smelting Furnace, developed to commercial application at Avonmouth in Britain in the early 1950s. Initial work was done at Port Pirie in 1951.

Port Pirie People

Dr G.K. Williams made a great contribution to the advancement of lead metallurgy through the invention, development and application of continuous and full refining of silver-lead bullion. This process remains

Casting pigs of refined lead at Port Pirie, about 1915.

The lead smelter and refinery at Port Pirie in 1980. The 205 metre chimney stack is the second tallest in Australia.

Frank Green, who spent his whole working life with The Broken Hill Associated Smelters Pty Ltd, was for 16 years Manager of the Port Pirie Smelters. He was the President of The Institute in 1963.

unique in the world and earned him the Institute Medal in 1942. The University of Melbourne named their metallurgical research centre in his honour.

The BHAS operations are much more than a production plant. They have been, and remain, a great centre of innovation. Two Works Managers: Dr G. K. Williams and Frank A. Green, were both metallurgists. Frank Green was for many years a Councillor of The Institute and served as President in 1963.

Frank Green was succeeded by John Funnell, an economist and accountant, who remained as Works Manager until he was transferred to Melbourne as a CRA Group Executive. John Funnell's contribution to the industry and The Institute was acknowledged by his election as an Honorary Fellow in 1989.

Health and the Environment

Equally important as the technological developments at Port Pirie have been the improvements in working conditions, health care of the employees and the surrounding environment.

George Dey, who qualified in mining and metallurgy from the famous Otago School of Mines in New Zealand, was appointed in 1925 to control health and hygiene at BHAS. His philosophy was that at 'the end of every shift, every man should be able to leave the works as fit and as well as he began that shift'.

In 1925, there were 261 cases of lead poisoning at the smelters, 30 in 1930, and only two in 1935. Today cases of lead poisoning are rare. In 1951 Dey was awarded Honorary Membership of The AusIMM.

Keast, who often told of his desire to see 'two trees grow where one grew before', enthusiastically supported the regeneration programme initiated during his years at Broken Hill and developed by the self-taught botanist, Albert Morris, who was able to show that the encroaching sands could be stopped and that grasses, shrubs and trees could be grown.

Keast's involvement with Port Pirie began in the late 1940s, when he became General Manager of BHAS, with responsibility for the smelters. At that time there was an increase in staff at Port Pirie, especially those engaged in research, and homes were needed. With his experience at Broken Hill and how conditions there had been improved, he ensured that the architect was briefed to include space for trees. The area became virtually tree-covered, with homes nestling in an environment as appealing as many city suburbs.

Hand charging a lead blast furnace at Port Pirie, about 1915.

Another Industry at Port Pirie

In 1944, during the Second World War, the British Government asked that Australia search for uranium. The reasons, of course, were not then widely known.

Two deposits in South Australia held promise as potential sources: Mount Painter and Radium Hill, both of which had been worked on a small scale for radium. The initial efforts were directed to Mount Painter, but the results were inconclusive and the British and Commonwealth Governments withdrew their financial assistance.

Tipping lead blast furnace slag, Port Pirie, about 1915. Occupational health and safety rules about dust were less rigid at that time.

The South Australian Government, under Sir Thomas Playford, GCMG, supported further work at Radium Hill and by 1949, under the direction of Sir Ben Dickinson, the Director of Mines, adequate reserves were proved. The Combined Development Agency wrote a contract which enabled development to proceed. Full scale mining at Radium Hill began in 1954 and continued until 1961. An acid leach and ion exchange plant for the production of uranium oxide was built at Port Pirie and operated between 1955 and 1962. Before it closed, a pilot plant recovered a quantity of scandium and yttrium. The uranium plant was later acquired by a private company for recovering a range of rare earths from the

The Port Pirie plant which extracted uranium from the concentrates railed from the Radium Hill mine between 1954 and 1961.

tailings. It was technically successful, but uneconomic and soon closed.

In 1990, the newly formed SX Holdings Limited proposed to set up a rare earths processing plant to retreat the Radium Hill tailings and recover a range of both light and heavy rare earths, with various applications within emerging technology fields such as magnetostrictive rods and certain magnet alloys.

Treatment of wastes was not new to BHAS. For years the large residue dump which once dominated the landscape had been steadily retreated for its zinc content, which for many years could not be economically recovered. Again new technology, much of it developed locally, aided in the solution of a complex metallurgical problem.

Future Outlook

The Port Pirie smelters, now a part of Pasminco Ltd, have had a long and respected history in the invention, development and refinement of a variety of intricate metallurgical process and associated technical developments, which bring great credit to the many who have been involved. They have served not only Broken Hill and Port Pirie but the nation well, and have given technical leadership to the world. This contribution, so important in the past, is continuing today.

I am indebted to John Dew and Gilbert Ralph, MBE, for their invaluable assistance in the preparation of this paper, which was presented in Port Pirie on 13 August 1990.

Iron Ore and Steel from Whyalla

If we relied only on media stories, we could be forgiven for thinking that Australia is full of wetlands, rainforests and mighty rivers running through pristine wilderness. The predominance of harsh, arid climate and the isolation of the people who live in these areas, which have been so important to shaping the nation, has somehow been lost from today's popular press.

Whyalla typifies all that has been forgotten. The story of Whyalla is one of hard-won change, of continual striving, of incremental improvement: in economic activity, in coping with the harshness of the environment and overcoming isolation. It is a story with its share of setbacks as well as successes and triumphs.

I am mindful of the parallels with someone who was very much involved in early developments at Whyalla and Iron Knob. I speak, of course, of Essington Lewis, CH, who became Managing Director and Chairman of The Broken Hill Proprietary Company Limited (BHP). He too, strove continually for ways to improve, took on one challenge after another and throughout his life knew much hard work. His personal style was not flamboyant, yet we in the industry all know about his importance to Australia. Much the same is true of Whyalla.

Discovery and Early Development of Iron Ore

The explorer Edward John Eyre is credited with first recognising the presence of iron ore in the area. His journal entries of 18 September 1840 are thought to refer to Iron Baron. He apparently passed, but did not climb, Iron Knob. By 1854, Iron Knob was named and recognised as a large iron ore deposit. Iron ore was of no particular value then, with no domestic steel industry. Nothing much happened for another half century.

It was gold and silver that made hearts beat faster and eyes shine more brightly in 19th century Australia. In South Australia, copper was being mined with considerable rewards, but when iron ore leases were first made available in the Middleback Ranges in the 1880s, there was 'nothing in the nature of a stampede to get [them]'.

In the 1890s leases were taken out over parts of the deposit. The quest was for another bonanza like Broken Hill, where Charles Rasp had come upon a huge silver-lead field, but there was no such luck in the Middlebacks. With hopes of finding gold or silver dashed, the leases were allowed to lapse.

Development in the Whyalla area finally began in a roundabout way, with BHP's decision to move its silver-lead smelter to Port Pirie, on the opposite side of Spencer Gulf from Hummock Hill, the settlement that

Iron Knob, about 1911, before the mining of iron ore for use at Newcastle steelworks.

Iron Knob in 1915, soon after the start of mining of iron ore for steel making.

was to become Whyalla. The smelter created a demand for iron ore, to be used as a flux in the smelting process.

In 1897 R. D. Hale pegged claims over Iron Knob and Iron Monarch for BHP. Another man, Burchett, who had pegged and prospected the Knob, concluded they were large mountains of iron ore, and he was right. The deposits at Iron Knob and in the Middleback Ranges were the largest then known in the whole country and 'as rich a deposit of ore as any in the world'. At the time this was of little importance, BHP's requirements for flux being quite modest.

To meet the needs of the Port Pirie smelter, a quarry was opened up at Iron Knob. In some ways this early mining activity resembled the public's more romantic view of mining, of the fossicker toiling away. The men involved doubtless had a less rosy view of it all.

Drilling in those days was a matter of hammer and tap, two men wielding 14-pound hammers to tap the drill, a third holding it in position. Bill Kittel was one of the original ore loading contractors at Iron Knob. His solo record of loading 14 trucks with ore in one shift was never bettered. With each truck taking five tons of ore this amounted to 70 tons, although Bill was heard to observe wryly, 'Those damn trucks used to hold nearer six tons than five'.

At first the ore was taken by bullock wagon to Port Augusta and railed to Port Pirie. The teams took two days to reach Port Augusta, another two to return. The cost of transport was out of all proportion to the value of the ore.

All this changed when BHP built a tramway from Iron Knob to Hummock Hill and shipped the ore across the Gulf to Port Pirie, loading it on the 390-metre jetty it built. The tramway required the approval of the South Australian Government, although (perhaps fortunately) there was no requirement for an Environmental Impact Study. The Hummock Hill to Iron Knob Tramways Act 1900 was introduced as a Private Member's Bill by the Hon. John Darling (later Chairman of BHP) and considered by a Select Committee chaired by none other than Essington Lewis's father. The 53-kilometre tramway was completed in 1901. The Act's various regulatory measures included a speed limit of 40 kilometres per hour, though it is hard to say how this was policed.

In this way, a modicum of infrastructure began to be built up in the region. Developments also went ahead at Iron Knob. In one of Australia's first experiments in company housing, BHP built houses there from 1905.

Essington Lewis, CH, first visited the region in 1906, apparently to investigate rumours that gold had been found. In those days Hummock Hill had 'hardly a tree which gave enough shade at midday to cover a sheepdog'.

With an annual rainfall of only 225 millimetres the region is indeed a parched one. In the very early days, water was shipped to Hummock Hill from Port Pirie. This was satisfactory only while the population remained small, and lack of water continued to be a significant constraint for quite some time.

The first load of iron ore for Newcastle leaves Whyalla in Emerald Wings, *January 1915. The ship was later renamed* Iron Baron.

A Turning Point: BHP's Entry Into Steel Making

BHP's decision to use the iron ore deposits of the region as a basis for a steel industry marked a major turning point. When an experimental batch of pig iron was made at Port Pirie in 1905, the quality was good, indicating no impurities in the ore. Still, it was not until 1911 that BHP decided to establish a steel industry.

Various reasons have been given for the decision: the Board's recognition that its great mine at Broken Hill was well past its peak; soaring demand for steel in Australia, coupled with the young Federal Government's willingness to protect domestic producers; declining base metal prices; and Guillaume Delprat's conviction that the steel industry was the only sure course for the future. Whatever the reasons, the decision carried with it the seeds of Whyalla's development into a centre of great importance to the industry. It ushered in a period of rapid growth and brought to the area a man who would later be seen as one of Australia's industrial giants.

Stationed at Port Pirie, Essington Lewis, CH, saw to the expansion necessary at Whyalla and Iron Knob to meet the needs of a steelworks. This involved opening a new quarry, housing an expanded workforce, upgrading the tramway and jetty, and installing a conveyor belt and bulk-loading equipment. The steelworks development work was completed late in 1914, several months after the outbreak of the First World War. The first shipment of iron ore, bound for the Newcastle steelworks, was loaded in January 1915. The Institute's first visit to Whyalla was in 1913 when all this activity was under way. In the following year, the name Hummock Hill changed to Whyalla.

Tramway incline, Iron Knob iron ore quarry, about 1915.

Further expansion took place during and just after the War as demand increased. The tramway's original 40 pounds to the yard rail was replaced by 80 pounds to the yard rail. Introduction of new, larger locomotives in 1915 to haul 2,700 tons of ore daily was without incident, once a large box had been provided for one driver of smaller stature, who needed it to stand on so he could reach the controls. The ore handling plant at Whyalla was upgraded to a capacity of 1,000 tons per hour and in 1918 pneumatic drills replaced hand drills at Iron Monarch.

By 1920 Whyalla's population though still only 1,000 had increased ten-fold since the early days of operation. It was still isolated and truly a company town, with services and amenities (such as they were) provided by BHP. A community hall, built in 1905, was used for church services as well as other community activities. Further additions to the amenities were somewhat delayed. The story goes that Essington Lewis and Guillaume Delprat, CBE, could not agree on whether an institute or a church should be built next. Lewis favoured the former, but Delprat, a deeply religious man, argued for a church. The argument was solved in 1920, when both were built. Two years later a desalination plant, capable of handling 1,000 gallons of seawater per hour, solved the water problem for a time.

In the 1920s demand for iron ore via Whyalla was on the increase, particularly after it was shipped to Port Kembla for the first time, in

Iron ore train between Iron Knob and Whyalla, about 1928.

1927. Demand had gone beyond the scope of hand loading, so an electrically operated plant was installed at Iron Knob that year. Power was supplied from Whyalla by a high tension transmission line of 33,000 volts, the first all-Australian high tension line built. In the 1930s, during the Depression, Iron Prince and Iron Baron were opened up, adding to the supply from Iron Knob. In 1935 ore haulage and shovel operations were electrified and the ten millionth ton of iron ore was shipped from Whyalla.

During this period there was a concerted effort to improve the town. When Essington Lewis, CH, first visited Whyalla it had little in the way of vegetation. When he became General Manager, one of his first steps was to encourage the growth of seedlings and planting of trees at Whyalla and Iron Knob. Lack of water hampered progress. In 1932 Albert Morris, chief assayer for Central Mine at Broken Hill and the keen amateur botanist who was prominent in creating the regeneration area in Broken Hill, visited the area and laid down a plan for tree growing. Lewis, visiting California in 1934, spotted a tree he thought would flourish in the arid country around Whyalla and from cuttings he imported the Athel pine became a familiar sight around the area.

Following Essington Lewis's visit to Japan in 1934, iron ore production increased rapidly. Lewis was convinced the Japanese posed a serious threat and that Australia should prepare itself for war. Output from the steelworks rose rapidly between 1934 and 1941, until BHP could make more steel than Australia needed.

A Blast Furnace for Whyalla

The next chapter in the story of Whyalla began in 1937, when BHP decided to set up a blast furnace. This was in part because of State Government pressure, as the South Australian Premier, Richard Butler, had pressed the company to establish a steelworks. The implied threat was that new iron ore leases, due in 1944, might not be granted. BHP complied, in part because of Essington Lewis's conviction that there would be another war. In 1937 the South Australian Government granted a 50-year extension to the leases, in return for BHP building a blast furnace and harbour at Whyalla and promising to build a steelworks. Dredging of the harbour began in 1938 and the blast furnace was blown in three years later.

The Whyalla blast furnace, 1948. It and the adjacent shipyard were built during war time following an agreement with the S.A. Government.

Before these new developments began, Whyalla still had only 1,400 people. But with the coming of a blast furnace this would grow rapidly, exacerbating the problem which had always plagued Whyalla, lack of water.

By 1940 the population had reached 4,300. A radical solution to the water supply problem was proposed: to pipe water 460 kilometres from the Murray River at Morgan. Work began on this ambitious project in 1941, on delivery of the first concrete-lined steel pipes. In spite of wartime interruptions, the first water from the Murray flowed into Whyalla only three years later.

Other amenities were also provided: a hospital in 1940, a maternity wing in 1943. The BHP-owned dairy, set up in 1935, was moved to a larger site in 1941, at the same time as the abattoirs. By 1945 the dairy was producing almost 1,400 litres of milk per day for a town which had grown to 7,500 people.

Related Mineral Developments

The blast furnace encouraged other developments. The steel industry in the east had relied upon deposits in Tasmania for its limestone needs, but as early as 1925 BHP had expressed interest in finding high-grade limestone in South Australia, within reach of the coast. BHP's Superintendent of Mines and Quarries, Frank Hockey, inspected a deposit at Rapid Bay and later reported that the deposit was 'by far the most attractive of the 30 or so I have inspected this year'.

The quality of life at Whyalla and Iron Knob gradually improved. Swimming pool at Iron Knob, about 1955.

Hockey set about gaining access to the deposit once the Newcastle steelworks reported that five of the six limestone samples he had sent them were suitable for use. An agreement was signed with the landowners in 1926 but 14 years passed before any mining took place. In 1940, work finally went ahead, a loading jetty and quarry plant having been sanctioned by the Rapid Bay Indenture Act of 1937.

In September 1942, the first shipment of over 5,000 tons of limestone reached Whyalla, where the blast furnace had been completed. Work had begun at Rapid Bay in 1940, with 24 men. By the early 1950s the plant was producing 40,000 tons of screenings monthly with 44 in the workforce. When BHP sold the quarry, in 1981, it had produced 14,000,000 tonnes of limestone in 40 years of operation.

Dolomite quarries were opened up at Ardrossan in 1950, to supply refractory and fettling material for Newcastle and Port Kembla. Dolomite was later used as a flux, with the introduction of Basic Oxygen Steelmaking at Whyalla.

In 1945, the South Australian Parliament passed the Whyalla Town Commission Act, ending BHP's municipal government in the town. The Town Commission was a unique experiment in local government, comprising seven members: three nominated by BHP, three by ratepayers and one by the Government as Chairman and Chief Executive Officer. This arrangement remained until 1970, when it was replaced by a fully-elected local government.

Establishment of Steelworks

The last great burst of growth in Whyalla took place after BHP's 1958 decision to construct an integrated steelworks. Again, the South Australian Government had pressed for this, in return for new iron ore leases. The population trebled in just a few years and Whyalla was proclaimed a city in 1961. To cope with the influx the South Australian Housing Trust was, in that year, building brick homes at the rate of one a day. This rapid growth saw Whyalla become the State's second-largest city.

The plant opened in 1965, with a new blast furnace and the first coke ovens installed in 1968. By 1972 the standard gauge railway connected Whyalla with Port Augusta, making further inroads into her historic isolation.

Whyalla has had her share of disappointments. Shipbuilding operations ceased in 1978, a total of 66 vessels having been completed. And naturally Whyalla was not isolated from the great upheavals in the steel industry during the early 1980s.

Whyalla Today

Whyalla emerged the stronger from the dark days of the early 1980s. The population dropped somewhat, and the workforce at the steelworks was reduced, but Whyalla's steel industry became world competitive.

Much was achieved through greater co-operation. Agreements between management and unions in Whyalla have served as something of a model for other parts of the country. Everyone came away with a better understanding of what it means to be competitive. Further developments were on a sound footing.

Whyalla has had a part in mining developments elsewhere, notably in those that depend for their economic viability on heavy-duty railways. In 1979 the city was chosen as the major rail rolling centre in Australia. Three years later, plant to head-harden rails for heavy-duty railways was

Ampol tanker P.J. Adams *about to be launched at Whyalla in 1962.*

commissioned, with a second plant added in 1985. Rail head-hardening technology developed for Whyalla was sold to a United States steelmaker, and this achievement was recognised by an Australian Export Award.

More recently BHP approved the installation of a high technology continuous slab caster. Planned to be commissioned in 1992, this facility would further strengthen Whyalla's world competitive position.

Developments were not restricted to the steelworks. In 1984 Santos Limited opened a fractionation plant at Port Bonython, and in 1990 BHP opened a new mine at Iron Duke. Renewed confidence in Whyalla was also reflected in the city itself, in such things as development of the foreshore and the Hummock Hill scenic lookout.

Perhaps most encouraging were the efforts of The Institute's younger members in Whyalla. Much of the responsibility for recent work at both Whyalla and Iron Duke has been borne by some younger shoulders.

Whyalla's mineral heritage remains of paramount importance. The ore which has sustained the region for so long continues to be the

Shunting two 250-tonne capacity torpedo hot metal ladles of molten iron from the blast furnace at Whyalla to the basic oxygen steelmaking (BOS) plant, 1970s.

Rolling mill at Whyalla.

foundation of her well-being. Production at the iron mountains – Iron Knob and Iron Monarch – continues after almost a century, though it is down from its peak. Some areas, having given up their ore, have been rehabilitated to resemble the appearance of the surrounding hills. Development of the Iron Duke deposit ensured the Whyalla steelworks' needs for 20 years. Other deposits, and large quantities of lower grade ore remain, perhaps to be mined in the future.

Professor Eric Rudd, AC, a man who knows this region well, recently made some estimates of the value of the iron ore produced there. Total production to the end of 1988 was approximately 190,00,000 tonnes. With the average price for export at $23 per tonne, this values the ore at more than $4 billion. About $40,000,000 had been paid in royalties.

The great efforts made to establish and maintain the mining and steelmaking operations in Whyalla and the contributions made to Australian industry have gone and continue to go unsung and unknown to most Australians. I began by suggesting a parallel with Essington Lewis, CH. His story is not well known by the general public, but his contribution was tremendous. He was Institute President in 1935 and received the Institute Medal in 1940.

One wonders where we would be without him. The same applies to Whyalla.

I gratefully acknowledge the major contribution of Vicki Lukins, and the assistance from Jeff Lomas of BHP Steel in the preparation of this paper, which was presented in Whyalla on 22 March 1990.

Victoria

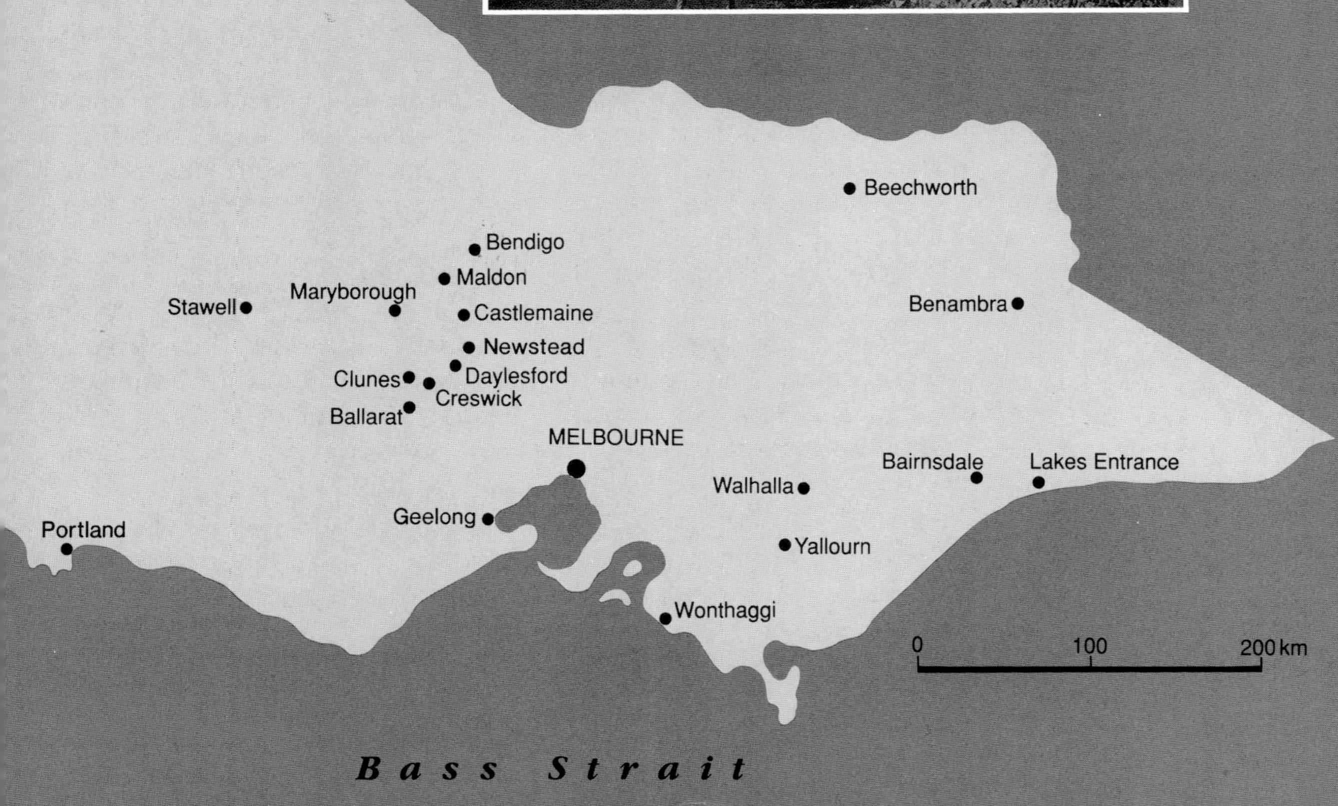
Beechworth
Bendigo
Maldon
Maryborough
Stawell
Castlemaine
Benambra
Newstead
Daylesford
Clunes
Creswick
Ballarat
MELBOURNE
Bairnsdale
Lakes Entrance
Walhalla
Geelong
Portland
Yallourn
Wonthaggi
0
100
200 km
Bass Strait

Melbourne's Entrepreneurial Role

It is a happy coincidence that this opportunity for me to speak to the Melbourne Branch occurs on the eve of Saint Barbara's day. I started my mining life as a student at the Clausthal Mining Academy in the Harz Mountains in Germany in 1946 and worked in the lead-zinc mines in the Harz and the coal mines in the Ruhr before I came to Australia.

Mining traditions in Germany go back more than 1,000 years and Saint Barbara is very much a part of them. She is the patron saint of miners and cannoniers. It is not clear why these two professions are taken together by the heavenly authorities, unless it is that both involve making loud bangs and living somewhat dangerously.

The Role of Entrepreneurs

Finding orebodies and developing mines and mineral activities does not originate in politics, or ideology, or laws, or regulations. It arises from the ideas of people and their enterprise, knowledge, and willingness to exert themselves and take risks. Our mineral heritage is very much the result of the ideas and actions and successes and failures of our predecessors.

It is the heritage of entrepeneurship in Melbourne which I talk about today, not the kind of entrepeneurship which has given the word a bad flavour in recent years, but real entrepeneurship of creating real value which benefits the community. I will have to restrict myself to a small number of people and events. Those mentioned are given as prominent examples, in the knowledge that many others deserve similar acknowledgment.

In the early gold rush days of the 1850s, people flowed into Melbourne so rapidly that tent cities, such as this one at Emerald Hill, grew up to house them until they made their way to the Goldfields or found work in town.

The Gold Rush Era

In our early gold rush days Melbourne was the beneficiary, rather than the leader, of mining developments.

Wealth from the Central Victorian Goldfields led to the birth of the provincial cities of Ballarat and Bendigo, and the benefits flowed to Melbourne and Geelong. In 20 years, gold transformed Melbourne from a struggling, impoverished rural-based town to a city with fine public facilities and buildings such as the University, Public Library, Court House, Parliament House, Town Hall, and theatres. Its wide streets, attractive public gardens, and elegant mansions led it to be called Marvellous Melbourne.

Corporations and Financiers

Entrepreneurs entered the scene when the alluvial gold ran out, for mining reef gold required more capital and organisation than individuals could provide.

The main metal mining centres in Australia at the start of the 20th century included Kalgoorlie, Wallaroo-Moonta, Mount Morgan, Mount Lyell, Ballarat, Bendigo and Broken Hill.

Kalgoorlie was mostly controlled from London, Wallaroo and Moonta from Adelaide, and Mount Morgan from Sydney. The Victorian Goldfields were for the most part controlled from Ballarat and Bendigo by people like George Lansell.

But the field that was to play the greatest role in Australia's economic development, Broken Hill, was mainly directed from Melbourne, and it was through Melbourne that the greater part of the wealth from Broken Hill was redirected to other entrepreneurial developments. The major examples of this are The Broken Hill Proprietary Company Ltd, or BHP as it is usually known, and the former Collins House group of companies.

Melbourne's prosperity, largely based on gold, enabled it to build such churches as Collins Street Independent and Scot's.

The Broken Hill Proprietary Company Limited

Let me quote from BHP's centenary publication, *How It All Began*:

> It was a time of beginnings, the 1880s in Australia. There was world-wide interest in the six colonies, for good reason.
>
> The earlier gold discoveries had started a pattern of investment and industry development. Land under cultivation doubled during the 1880s, Farrer began his wheat experiments and the Chaffey brothers started irrigation schemes. By then, nearly two-thirds of the population of 2,500,000 had been born in the country.
>
> Sydney and Melbourne were linked by rail in 1883 and Melbourne had its first cable tram in 1885; there were already double-decker steam trams in Sydney...

Jamieson Shaft and the open-cut mine, in the early days of Broken Hill.

Coke ovens and plant, Australian Iron & Steel Ltd, Port Kembla, 1938. The Port Kembla plant came under Melbourne influence with the BHP takeover of AI&S in 1935.

... And on Sunday, 21 June, 1885, one William Jamieson set out from Silverton in the far west of New South Wales to arrange for the flotation of a new mining company on Australian stock exchanges. That company was eventually incorporated in Melbourne on 13 August, 1885; it was The Broken Hill Proprietary Company Limited.

BHP held the central part of the Broken Hill line of lode. It started to mine the extremely high grade silver-lead of the enriched ore zones and paid very handsome dividends to its shareholders.

William Jamieson in 1885 arranged the issue of 2,000 shares at £9 each. Of the proceeds £15,000 went into mine development, and £3,000 was divided among the 14 syndicate members who had also been allotted 1,000 shares each. By 1891, dividends paid had exceeded £1,000,000.

In 1899 the General Manager, Alexander Stewart, resigned and Guillaume Delprat, CBE, replaced him. Delprat had a significant impact on BHP and the industry. He was President of The Institute in 1906 and in 1935 was the first recipient of the Institute Medal.

In 1911 the company faced a major decision. With 80 per cent of the ore reserves within its leases already mined, there was the possibility of continuing as a custom smelter (smelting ore for other companies) at Port Pirie when the Broken Hill ore ran out. But Delprat urged his Directors to consider getting into steel-making.

The idea was to use the large known ironstone resources of the Middleback Ranges on the west side of Spencer Gulf, which since 1901 had been a source of flux for the Port Pirie lead smelters. BHP had set up its lead smelters in Port Pirie in the 1890s, mainly because the site offered economies of transport over smelting in Broken Hill.

In June 1911 the Board sent Delprat to the United States to study steel production. There he recruited David Baker, a Philadelphia consulting engineer. Baker and Delprat advocated setting up of a steelworks at Newcastle where coking coal was available, using iron ore from the Middleback Ranges.

Hand mining at the original Iron Knob quarry, in the Middleback Ranges of South Australia, 1915. BHP's Melbourne-based management saw the opportunity for an Australian steel industry using South Australian iron ore and New South Wales coal. The steelworks at Newcastle started production in 1915.

The Newcastle steelworks opened in 1915, exactly four years after the Board decided to send Delprat to the United States. This speed could not be repeated today, with its requirements of lengthy studies, enquiries, public consultations, and a plethora of approvals. It is a good question as to whether it would be even possible to start such an industry at all today, let alone in four years, when anything proposed is opposed by somebody.

It would be impossible for me to describe the contributions made by the hundreds of enterprising people who shaped BHP into the company it is today. I will restrict my comments mainly to just one person who serves as a prominent example.

Essington Lewis, CH, who had studied mining engineering at the South Australian School of Mines, joined BHP in 1904 at Broken Hill, where at first he worked underground. In 1905 he was sent to the smelters at Port Pirie, with responsibility for Iron Knob. By the time steel production began at Newcastle in 1915, Lewis was in charge of materials from the iron ore and limestone quarries and by 1917 he was in charge of all construction work at Newcastle. When Delprat left, in 1919, Lewis was named as Acting General Manager, 38 years old, easily the youngest of the steelworks' management team. In 1921 he became General Manager of BHP and in 1926 Managing Director.

Essington Lewis, CH, in 1921, the year he became General Manager of BHP.

Lewis led the company into coal, coke, rope, wire, nails, tubes, sheets and a variety of steel products either through formation, takeover or amalgamation. Brand names such as Rylands, Titan, Lysaghts, AI&S, Hoskins, Comsteel, Rheem and Wiltshire became part of BHP.

Lewis was President of The Institute in 1935, and was awarded The Institute Medal in 1940 in recognition of his part in the development of the steel industry and for his service to the nation.

Directors of Broken Hill Associated Smelters Pty Ltd, most of whom were Melbourne based, at the Directors' Lodge, Port Pirie, in 1926. Back row: Sir Alexander Stewart, John Govett, Sir Colin Fraser, H. St. J. Somerset. Front row: Oliver Lyttleton (later Lord Chandos), W.S. Robinson, W.L. Baillieu, Sir Robert Horne.

Wirraway fighters in production at Commonwealth Aircraft Corporation Pty Ltd, Melbourne. Essington Lewis and other mining leaders provided the original drive that led to the development of Australia's own aircraft industry before and during the Second World War.

Lewis anticipated the Second World War and with W. S. Robinson, then Managing Director of the Zinc Corporation Ltd, and Sir Laurence Hartnett, Managing Director of General Motors-Holdens Ltd, formed the Commonwealth Aircraft Corporation Pty Ltd (CAC) in October 1936. Lewis became Managing Director of CAC and by January 1939 the first Wirraway was in the air. In 1938 Lewis was invited by the Commonwealth Government to chair a panel to advise on the mobilisation of private industry in the event of war. This led to his appointment as Director General of Munitions when war came.

At the end of the war, Lewis returned to BHP to oversee its reconstruction and redevelopment. In 1950, at the age of 69, he became Chairman for two years, after which he remained on the Board until his death in 1961 at the age of eighty. His successors at BHP have included other famous names, such as Norman Jones, CMG, Sir Colin Syme, AK, Sir Ian McLennan, KCMG, KBE, (Institute President in 1951, 1957 and 1972 and Medallist in 1959) and Sir James McNeill, CBE. It was under their leadership that further major developments took place.

BHP House, Melbourne from 1972 until 1991. BHP was incorporated in Victoria and has always had its headquarters in Melbourne.

In the 1960s BHP, with Esso Exploration and Production Australia Inc. (and the help of Lewis Weeks), discovered oil and gas in Bass Strait, and involved itself in the iron ore development at Mount Newman in Western Australia. In the 1970s it took up interests in coal in Queensland and in 1984 it acquired Utah International Inc. and its diverse worldwide interests.

From its origins in Broken Hill and the building of a modest steelworks at Newcastle, BHP has grown into a diversified international enterprise based on minerals, petroleum and steel. In 1990 it had a market value of about $15 billion, more than 50,000 employees, and an annual turnover of $14 billion. BHP has been Australia's largest company by market capitalisation since 1967.

The Collins House group

By contrast with Essington Lewis, CH, who was an entrepreneur working within an already established corporation, let me outline the involvement of entrepreneurs who were the promoters of new corporations and enterprises.

One such person was W. L. (Big Bill) Baillieu. From a start in the real estate business, he moved into full-time financial activity. He worked closely with his brothers in stockbroking and formed E. L. & C. Baillieu in 1894. This provided opportunities to finance and acquire interests in mining enterprises in Victoria and Broken Hill. It also provided valuable contacts in London's financial markets.

In 1902 one of the brothers transferred to the London Stock Exchange, then the world's most influential. This established a relationship with mining brokers Lionel Robinson, Francis Govett, and William Clark. (Lionel was an elder brother of the well-known London-based mining entrepreneur W. S. Robinson). Sustained base metal price rises led Robinson and Clark early in 1905 to acquire equity in North Broken Hill, Junction North, and the Block 14 companies at Broken Hill, as diversification from their Western Australian gold interests.

This so-called Anglo-Australian group in London formed a number of management syndicates with other London financiers interested in mining ventures.

The first was the Broken Hill Syndicate Ltd, registered in 1905 with the modest capital of £5,250, but with access to much larger resources. It formed BHSB Co. Ltd (Broken Hill South Blocks) through the acquisition of options over a number of Mineral Lease blocks, including potential ore and zinc rich tailings stockpiles. It also secured the leases at the south end of the Broken Hill field, which were developed and mined by the Zinc Corporation Ltd and New Broken Hill Consolidated Ltd from the late 1930s.

The second of these syndicates, the Hill Syndicate, included the mining consultants Bewick, Moreing & Co, plus an expert geologist and a mining engineer. The Hill Syndicate was instrumental in securing further large tonnages of zinc-bearing tailings at Broken Hill. These were treated by the Zinc Corporation and the highly-successful Amalgamated Zinc (de Bavay's) Ltd, using the emerging flotation technology.

The Zinc Corporation Ltd was formed in 1905. It struggled to survive financially as it tried to apply four different processes in five years to separate the zinc and other minerals. Its promoters, the Hill Syndicate, led by Lionel Robinson, were later joined by Govett, Bewick, Moreing's manager, Herbert Hoover (later to become President of the United States), and William Baillieu, who became Chairman in 1906.

Collins House, 360 Collins Street, Melbourne. Built in 1910, it became the headquarters of some of Australia's most successful mining companies and the non-mining subsidiaries they set up.

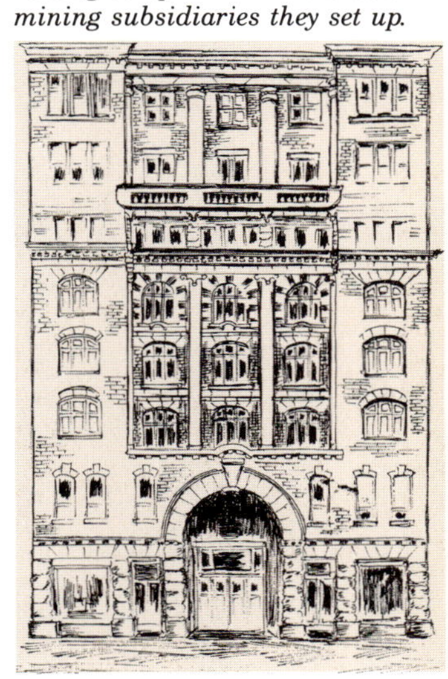

Paper machines at Associated Pulp & Paper Mills plant, Burnie, Tasmania. Melbourne-based mining companies reinvested profits, mainly from Broken Hill, in such ventures as paper manufacture.

W.S. Robinson became the Managing Director and the driving force behind the Zinc Corporation's growth through to the Second World War.

The Anglo-Australian group had its headquarters in Collins House in Melbourne, built by W. L. Baillieu and five of his brothers about 1910. The group drew together over the years a succession of leading engineers, technologists, and managers whose contribution to Australia has been outstanding. North Broken Hill Ltd, Broken Hill South Ltd, Zinc Corporation Ltd and Electrolytic Zinc Company of Australasia Ltd (EZ) were all members of the Collins House group, which also acquired from BHP its lead smelter at Port Pirie. Operated by Broken Hill Associated Smelters Ltd (BHAS), this became the world's largest lead smelter. It, the Broken Hill mines, and other lead-zinc operations, are now a part of Pasminco Ltd, which is owned by CRA Limited, North Broken Hill Peko Ltd and some public shareholders.

Sir Lindesay Clark, AC, KBE, CMG, MC, who played a prominent role in such major Australian mining companies as Western Mining Corporation Limited, Broken Hill South Limited and Alcoa of Australia Limited.

The emergence of the Collins House group from the early years of this century marked a new phase in the progress of Australian mining, from more-or-less independent companies based on individual ore deposits, to large integrated groups of world significance.

The combination of the ability to raise funds, good management, and some luck, made a major contribution to the economy and generated large surpluses for re-investment. This encouraged others to invest and resulted in further mineral discoveries and the creation of further enterprises in many parts of Australia, most, but not all, concerned with minerals.

During the First World War Amalgamated Zinc Ltd, which produced zinc concentrates at Broken Hill by the de Bavay flotation technique, financed EZ at Risdon, near Hobart. Sir Herbert Gepp, who was Manager of Amalgamated Zinc, became EZ's first Managing Director, based in Collins House. He was President of The Institute in 1924 and recipient of the Institute Medal in 1939.

Sir Colin Fraser, a New Zealander who had mining experience around the world before he attracted the attention of W.L. Baillieu and W.S. Robinson, became an important figure in the Collins House group of mining companies.

Sir Colin Fraser was a New Zealander who had graduated in geology and gained considerable experience around the world before attracting the attention of W. L. Baillieu and W. S. Robinson. He became an important figure in the Collins House group and was Chairman of a number of the companies based there.

Fraser and Robinson were amongst Australia's most distinguished mining leaders during the first half of this century. Both were honoured by The Institute – Fraser as President in 1928 and Robinson as recipient of the Institute Medal in 1949. Baillieu and Robinson were jointly awarded the Gold Medal of The Institute of Mining and Metallurgy in London in 1929 'in recognition of their services in the mineral development of the Empire', even though neither of them held any formal qualifications in related professions.

From Collins House in the two decades between the world wars also came initiatives and finance for establishing other industries, such as paint manufacturing in Sydney (British Australian Lead Manufacturers Ltd), expansion of the copper smelters and refinery at Port Kembla (Electrolytic Refining & Smelting Company of Australia Ltd),

manufacture of brass and copper sheets, rods and cables (Metal Manufactures Ltd), and a major pulp and paper mill at Burnie in Tasmania (Associated Pulp and Paper Mills Ltd).

The Zinc Corporation was later led by men such as A. J. Keast, CBE, (President of The Institute in 1946 and Medallist in 1963), and Sir Maurice Mawby, CBE, (President in 1953, 1954 and 1968, and Medallist in 1955). The company merged in 1949 with the Imperial Smelting Corporation Ltd to become Consolidated Zinc Corporation in Britain, with Consolidated Zinc Pty Ltd (CZP) as the Australian subsidiary. In July 1962 the Australian interests of The Rio Tinto Mining Company of Australia Ltd (RTMA) and CZP merged to form Conzinc Riotinto of Australia Ltd, now called CRA Limited. The CRA group has been instrumental in many major mineral developments since the 1960s, including Comalco Ltd, Hamersley Iron Ltd, Bougainville Copper Ltd and Argyle Diamonds Ltd. They moved from Collins House in the 1950s.

It was on the initiative of W. S. Robinson, who accurately forecast a rise in the price of gold in the 1930s, that Gold Mines of Australia Ltd was formed in 1930, funded from England and South Africa. From this group emerged Western Mining Corporation Ltd, Gold Mines of Kalgoorlie Ltd, Central Norseman Gold Corporation N.L., and other gold mining companies which undertook developments in Queensland, Victoria and Western Australia. Sir Colin Fraser was founding Chairman of these companies.

Sir Colin died in 1944 and was succeeded in most instances by Sir Walter Massy-Greene who, after a career in politics, rising to Deputy Prime Minister to Billy Hughes, embarked on an intensive business

Renewed interest in Western Australian gold in the 1930s led to the formation of various companies managed and financed through Melbourne. Gold Mines of Kalgoorlie Limited plant under construction at Oroya, 1937.

Transporting lead to the refinery at Port Pirie smelters. The enterprise that led to the establishment and development of the mineral processing industry at Port Pirie was Melbourne-based.

career, chairing more than 40 companies, mostly in Collins House. He was followed in the Western Mining group by Sir Lindesay Clark, AC, KBE, CMG, MC, under whose direction Western Mining diversified out of gold into other metals, which amongst other things led to the formation of Alcoa of Australia Ltd. Sir Lindesay was the President of The Institute in 1959 and was awarded The Institute Medal in 1962.

Sir Walter Massy-Greene, who chaired more than 40 companies, mostly based in Collins House.

Future Outlook

Five generations span most of Australia's mining history. In these five generations a number of leaders stand out as having been instrumental in the major developments in the Australian mineral industry. A number of these, and the corporations they led, were Melbourne based. People and corporations in other centres have, of course, also made a great contribution. My purpose in speaking about the role of Melbourne here today is not because of parochial prejudice, but in an endeavour to highlight the mineral heritage of this particular centre while speaking to the Melbourne Branch.

The need for leaders – people and corporations – to initiate and lead major new developments is as acute today as it was during the times we have reviewed. In this, members of The Institute will have a prominent role in the future, as they had in the past. Understanding the heritage of our predecessors is an important basis on which to build the future.

I am indebted to John Reynolds of Victorian Chamber of Mines and Gilbert Ralph, MBE, for their contributions to this paper, which was presented in Melbourne on 3 December 1990.

Victoria's Golden Wealth

Reports of payable gold discoveries in New South Wales in 1851 soon reached Victoria and created great excitement. People began to depart for goldfields north of the Murray River.

On the basis of 'if you can't beat them, join them', Victoria offered rewards for the discovery of a payable goldfield, in the hope of stemming the tide. Gold was discovered at Andersons Creek, near Warrandyte, east of Melbourne, by Michel and party on 16 July 1851, shortly after Victoria was proclaimed a separate colony. They received a reward of £1,000, which probably exceeded their own, and everyone else's, income from gold mined at Andersons Creek.

James William Esmond claimed to have found gold at what is now known as Clunes on or about 28 June 1851, a claim which was later challenged. Still, Esmond had brought to Victoria from California the knowledge, technology, and organisation for gold mining, and was granted £1,000 for being the first person in Victoria to mine gold economically.

During the next four years gold was found across Victoria from Stawell to Bendoc. As news of each new discovery reached the prospectors, they would hurry on to the next field. More and more people surged backwards and forwards across the colony in search of success. Three of the most significant gold discoveries were made within two months of the Clunes discovery; at Castlemaine, Ballarat and Bendigo.

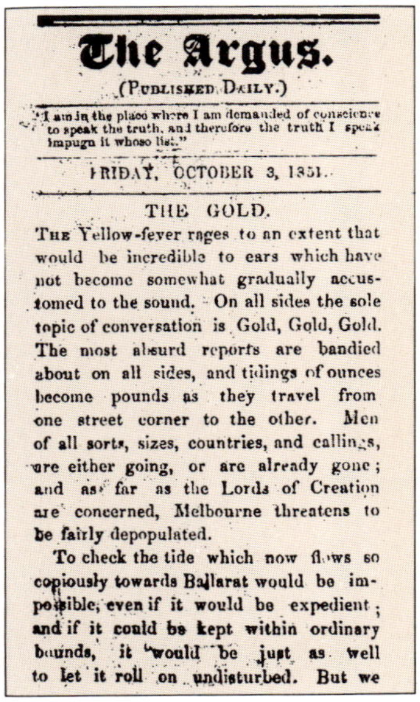

The Argus.

(Published Daily.)

"I am in the place where I am demanded of conscience to speak the truth, and therefore the truth I speak impugn it whoso list."

FRIDAY, OCTOBER 3, 1851.

THE GOLD.

The Yellow-fever rages to an extent that would be incredible to ears which have not become somewhat gradually accustomed to the sound. On all sides the sole topic of conversation is Gold, Gold, Gold. The most absurd reports are bandied about on all sides, and tidings of ounces become pounds as they travel from one street corner to the other. Men of all sorts, sizes, countries, and callings, are either going, or are already gone; and as far as the Lords of Creation are concerned, Melbourne threatens to be fairly depopulated.

To check the tide which now flows so copiously towards Ballarat would be impossible, even if it would be expedient; and if it could be kept within ordinary bounds, it would be just as well to let it roll on undisturbed. But we

Gold was discovered in Victoria in 1851. The Argus noted in early October that the 'Yellow-fever' raged to an incredible extent.

Gold Mining in the 1850s

To reach the Ballarat field from Melbourne, William Kelly in 1853 took a steamer to Geelong and then walked for two and a half days. He found Ballarat quite unlike California. Apart from the appearance of the landscape, the tents were flimsy, there was a large proportion of women and there were well-stocked stores.

He described his first descent down his friend's shaft. He was lowered on the rope with one foot in the sling and when reaching the bottom was impressed with the shaft excavation which was plumb vertical and perfectly round as if bored by an auger. Dry shafts were always round while wet shafts had to be slabbed with timber and so were rectangular.

If the shaft intersected gold-bearing clay, this was removed in a series of drives and cross cuts, often excavated by the miner in a prone position. Some shafts in 1853 were deeper than 30 metres. Parties of twelve men worked round the clock, four on each shift. Claims were about four metres square and Kelly records that from one claim, the Blacksmith's, £55,200 worth of gold was taken. Those who worked the claims were called diggers, a term of respect Australian soldiers were later to inherit.

Shaft sinking with a windlass, near Bendigo, 1852-3. Some miners were lucky enough to have their wives and families with them.

Steam engines began appearing in late 1854. Most were imported from the north of England and Scotland. Over 12 months would elapse between the ordering of machinery and its arrival. Their main use at Ballarat was to power mine dewatering pumps.

Deep lead mining was developed at Ballarat for mining waterlogged, buried river beds, not common outside Victoria. In time, 8,500,000 ounces of gold was recorded to have come from deep leads. At Bendigo there were no deep leads and as alluvial gold was exhausted miners either left for other fields or turned to the mining of gold-bearing quartz.

Cornish miners from South Australia, who had not found mining the soft and often dangerous alluvials to their liking, were in their element drilling and blasting quartz. The German, Theodore Ballerstedt, father of quartz mining in Bendigo, and others, had learned to mine and treat quartz in California. But many others were unskilled and learnt by trial and error.

Using a cradle to separate alluvial gold from pay-dirt. James Esmond brought knowledge of such equipment from California.

Quartz had to be crushed to liberate the fine particles of gold. Early steam engines on the Bendigo field were used to power the crushing equipment, as well as to dewater the mines. The most successful batteries were modelled on Californian designs and the gold was caught on Californian shaking tables.

As at Ballarat, there were women prospectors. On 1 September 1859 the mining correspondent of the *Bendigo Mercury* reported:

> In the course of our yesterday's tour we fell in with a lady fair who was adroitly and vigorously plying the pick on a little run or gutter, not far from Golden Square. We inquired, of course, into the prospect, and also if the gentlewoman had a Miner's Right. To the former portion of our interrogation she

replied that she had no doubt of making tucker – to the latter, that she despised the document in question, being a married woman, and having, therefore, instead of a Miner's Right – a Right to a Miner.

Bendigo had an air of permanence, even in 1854. There were many homes with stone chimneys, tended by housewives. The diggers had come to stay, and the towns of central Victoria were to become testimony to that.

Victoria very quickly overtook New South Wales as a gold producer, and for the next decade the average annual production exceeded 2,000,000 ounces. Between 1855 and 1863 Australia was the world's largest gold producer, accounting for up to 46 per cent of world production. Australian production peaked at 3,038,000 ounces in 1856, with 95 per cent of it derived from Victoria, where Ballarat and Bendigo were preeminent. In Australia only Kalgoorlie ever exceeded the cumulative production from each of these fields.

In some cases, women helping their husbands had zealous police demand that they pay the licence fee.

The gold won from the Victorian Goldfields brought enormous wealth and prosperity to the infant colony. It was the means by which the many fine public and private buildings were financed, not only in the Goldfields towns, but also in Melbourne and Geelong, which were the principal service centres.

Of all the Australian gold rushes, those to Victoria in the 1850s had the greatest impact on migration. People came from everywhere: Great Britain, Europe, America and China. Some came as merchants and tradesmen, but most came to dig for gold, in the hope of making a fortune. Few did. All but a few who could not resist the continuing lure of gold, in time went back to their former calling, or some other work, and stayed to build Australia into a nation.

The Companies

As the leads and reefs were followed deeper, and increased capital and technical skills were needed. It came mainly from local investors, sometimes miners, but more often storekeepers, tradesmen, and merchants, who recycled some of the wealth earlier won from gold. These associations were made possible by Ireland's Mining Companies Act in June 1858, and Pyke's Act two years later. The Acts covered the raising of capital, and the rights and obligations of shareholders, and were the beginning of company law in Australia.

Lansell's 180 Mine at Bendigo. By 1895 it was the deepest hole in the world, at 980 metres.

During the 1860s and 1870s much capital for Victoria's mines was raised in country stock exchanges. Exchanges at Ballarat and Bendigo were busier than Melbourne during booms, and places such as Happy Valley and Snake Gully, where today no buildings remain, had stock exchanges. Some companies also raised debt, which shareholders became liable to repay. To protect shareholders, the Victorian Parliament in November 1871 passed the No Liability Act for mining companies, limiting shareholders' liability to their investment in shares. This encouraged those with limited assets to buy shares and brought a resurgence of mining investment.

Typical gold digger of the early 1850s, with wide awake hat, bright shirt, moleskin trousers, boots and clay pipe.

Continued investment remained important because most mines operated on an accounting method similar to the Cornish Cost Book system, which limited development expenditure to the invested capital. Even when one of Victoria's last working gold mines, Wattle Gully, closed in 1969, it did so because profits were not allowed to be partly reinvested in development. Victoria's mine managers often had to resort to spiriting away funds to keep mines at work.

Most successful ventures were locally funded and managed. One exception was the Port Phillip Mine. The company was formed in London and sent a representative, Henry Bland, to Victoria to secure land at Fryerstown, near Castlemaine. His intentions inflamed the miners, who saw the leasing of large tracts of land and the introduction of machinery funded by foreign investors as a curb to their activities.

In time the company set up a mine at Clunes, a less important alluvial field, where it seems opposition was manageable, although only just. The mine opened in 1857 and operated until the late 1880s. At one stage the battery had 80 heads and was for a time the largest in the southern hemisphere. The company was responsible for bringing much useful technology to the colony, such as the self-feeding battery and copper plates with mercury troughs. Its miners were employed on a contract basis and this later developed in Victoria into tributing, related to the Cornish system, which remained a part of the Mines Act until recently.

The Technology

In the early 1860s, refractory ore was proving difficult. The Port Phillip company hired John Percy, lecturer in metallurgy at what became the Royal School of Mines, in London, to work on the problem. Percy, who was one of the world's leading metallurgists of last century, recommended roasting and then amalgamating the blanket material. This became standard practice in Victoria for pyritic ores.

Victorian mining was successful because it was not copying European theories and technologies. While English academics in the colony, such as McCoy, argued that gold would not occur at depth, the Magdala Mine at Stawell reached 720 metres by 1880. In the next decade Bendigo's mines were to go deeper and, with mines in Nevada, USA, were some of the deepest in the world at the time.

Most people walked to the diggings from the nearest port. For the Mount Alexander and Bendigo fields their starting point was Melbourne. For Ballarat, many chose to take a boat to Geelong and walk the easier grades from there.

In 1850 Melbourne had only a couple of small foundries, the only ones in Victoria. Ten years later most of the goldfields towns had a foundry. Of the substantial engineering works at Bendigo, Castlemaine and Ballarat, Thompsons of Castlemaine alone remains.

In 1884 the Berry No. 1 and Hepburn Estates deep-lead mines near Smeaton had Cornish beam engine pumps manufactured by the Union Foundry, Ballarat, with cylinders 1,800 millimetres in diameter and beams which weighed 50 tons. Victoria's foundries were, it seems, comparable with anything in Europe. Within one year of the arrival of Low's patent rock drill, made by Turner at Ipswich, England, in 1867, Ford's patent rock drill was designed and made in Victoria. It was a

better performer, although harder to maintain than imported drills such as the American National.

Mines in Victoria became centres of engineering. During the 1880s rockdrills were made in the machine shop of the North British Mine at Maldon. Such engineering capacity was most important for mines in remote areas. At Walhalla, for example, in 1868 one of Victoria's first safety cages was made and used at the Long Tunnel Gold Mining Company.

Ballarat School of Mines, the first of many such schools in Australia and New Zealand. It used a former Wesley Church (left) and former Circuit Court House as part of its premises.

Safety and Environmental Care

In the early 1870s Victoria's country newspapers began to lament the frequency of accidents and loss of life in the mines, and mining regulations were added to the Mining Act, modelled on English coal mine regulations. Mines inspectors, certified mine managers and licensed winder drivers soon became obligatory. Annual Reports of the Secretary of Mines during the 1870s and 1880s report with pride the decline in mining accidents and fatalities. Victoria's mining regulations became the basis for those in the other Australian colonies.

In the 1890s the Mines Department instigated the planting of trees on mined lands and this led to creation of the Forests Department. Some of the land so rehabilitated would in time become State Park. The Sludge Abatement Board, set up under the Mines Act, ensured that all lands being dredged would be rehabilitated and water courses became cleaner.

Many years before the modern environmental movement the mining industry was devising methods of returning the land to use after mining. One result can be seen in central Victoria at Newstead, where eight

When Ballarat settled down to steady mining, company promoters took over from diggers the task of funding mines. In the 1880s brokers and share buyers met regularly on street corners.

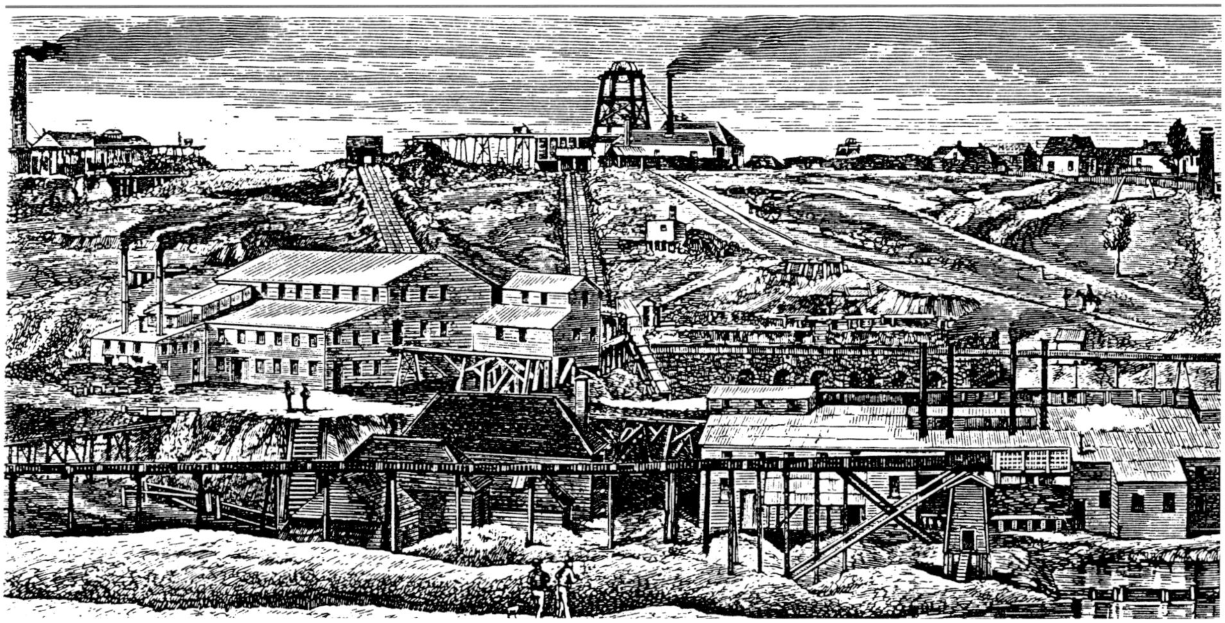

Port Phillip Company mine at Clunes. Formed in London, the company wanted land at Fryerstown, near Castlemaine, but local small-scale miners resisted to such an extent that it settled for a site at Clunes. The mine opened in 1857 and operated until the late 1880s.

kilometres of the Loddon River valley was dredged by the Victorian Gold Dredging Company N.L. in the 1930s and 1940s. Although much of their earth-moving equipment was impressed for the war effort, they still managed to return the dredged areas to pasture.

Contribution to the Community

Comparisons of Melbourne in 1839, 1854, and 1880 show the development of a sleepy village on the edge of the known world to a bustling frontier town and then to a sophisticated city.

The basis for the change was the wealth from gold. People flocked to Australia to escape famines, homelessness, poverty and Dickensian drudgery. Their journey presented uncertainty, hardship and considerable risk, but undaunted they came, to be part of something new, to be free and to improve themselves.

By the 1880s, after 100 years of European settlement and largely as a result of gold mining, Australia was the wealthiest per capita country in the world.

In central Victoria, almost every town owes its foundation to gold. The buildings in Victoria's gold mining towns – the churches, civic centres, hotels, mansions, and banks – testify not only to the affluence of the gold mining economy, but to the intention of the miners and their associates to make their homes there.

Among the many fine public buildings and mansions remaining from this era is 'Fortuna' at Bendigo. It began as the modest dwelling of Theodore Ballerstedt, who sold it and the adjoining mine to George Lansell in 1871 for £30,000 believing the mine to be exhausted. Lansell struck a bonanza and extended both the plant and his house to enormous proportions until the 200-head stamp battery and his three-storey mansion joined.

'Fortuna', at Bendigo is one of the Victorian Goldfields best known mansions.

Roads and railways were built from Melbourne. The railway to Bendigo, completed about 1858, is a testimony to the affluence and optimism of the time. It is high-quality double track, unlike most later Australian country railway lines.

But the great legacy of Victoria's gold is Australian society itself. During the 1840s there were many advocates of a free and democratic society in Victoria. Yet while the only incentive for settlers was the availability of land, Victoria was destined to become a pastoral society, where only land owners could share in government, control the economy, and enjoy full freedom.

That changed in 1854, when 200 miners stood at the Eureka Stockade. Peter Lalor, one of the leaders of the Eureka rebellion, later became Speaker of the Legislative Assembly. Many regard Eureka as the birth of democracy in Australia. While this may be an exaggerated view of the incident, as it was really a rebellion against discriminatory taxation, it was indeed an important event in the gradual emergence of a more equitable system.

Gold gave power to the idea of democracy. It enabled the creation of an egalitarian society where hard work and some luck, rather than parentage and privilege, were the means of success. Since that time all comers to Australia have had the opportunity to set up their own enterprise or to rise to the top of business, industry, or the professions.

Magdala mine at Stawell, one of those mines that disproved theories that gold would not occur at depth.

Poverty Mine at Tarnagulla, named not after its financial performance, but the fact that in 1862 a group of New Zealanders from Poverty Bay found an incredibly rich reef there.

The Institute

It was in Ballarat in July 1892 that the idea of forming the Australasian Institute of Mining Engineers (the fore-runner of The Australasian Institute of Mining and Metallurgy) was formally proposed by a Broken Hill delegate and adopted by the Annual Conference of the Mining Managers Association, held at the Ballarat School of Mines. The Institute was established in Adelaide in 1893.

Start of a new gold mine. Digging the shaft for the Nell Gwynne mine, Bendigo, in the 1930s.

The 1900s

The Victorian mining industry remained vigorous until the First World War. From then on, financial difficulties, technical problems and the restricted area of tenements for underground development, forced the shutdown of many mines.

A minor resurgence because of the gold price rise in the 1930s saw the start of some new operations in Victoria, notably at Bendigo, but with mixed success. With the Second World War, the decline of the industry resumed. Rising costs and the fixed gold price resulted in the closure of the last of Victoria's significant mining operations in the mid 1950s. During the 1960s and 1970s there were a few small intermittent operations, one of the most notable being the A1 Mine at Gaffneys Creek. In 1977, Victoria's gold production fell to a mere 354 ounces.

The 1970s saw a stream of environmental and social legislation, of worthy intent. The legislators, and more importantly, their advisers, seem to have had the view that mining was an industry neither needed nor wanted in Victoria. This appears to have been a view arising not out of a rational view of the facts, but out of ignorance and perhaps prejudice.

The new legislation, notably in terms of planning and later in planning and environment, combined with a convoluted and complexly evolved Mines Act, made it frustrating and very difficult to explore, let alone mine, in Victoria. Delays in granting approvals ran into years. Every step of the way seemed designed to be as negative as possible, and this was the main factor preventing Victoria from sharing in the great elevenfold increase in Australian gold production during the 1980s.

The Present

In 1990, Victoria had only three modest gold mines operating and minor production from a few tailings retreatment and small scale mining operations. In 1989, Victoria's metallic mineral production was less than 0.5 per cent of the Australian output, a sad commentary on a State which historically had produced 35 per cent of Australia's gold.

In 1990, gold was the only metal mined in Victoria. The copper, zinc, and silver project at Benambra, in the north-east, and the extensive heavy mineral deposits found near Horsham, in the Wimmera, showed that other minerals could be discovered there.

But Victoria has to amend its tedious and uncertain official procedures if it wants to attract mineral investment and development.

Logo adopted by the Australasian Institute of Mining Engineers on its formation in 1892. This body was the fore-runner of The Australasian Institute of Mining & Metallurgy.

Dredging is another way to separate alluvial gold. Victoria Gold Dredging Co. worked on the Loddon River flats near Newstead between 1938 and 1948.

Panning for gold at Sandhurst Town, near Bendigo. Genuine prospectors would never have tolerated such a curious audience.

It needs bold thinking by the State's political leaders to undo the damage caused by legislation and practices which have failed to understand that social and environmental objectives can only be met if there is continued investment and economic growth. Given a chance, the mineral industry has the potential to make a major contribution to Victoria's future. Whether it will be given this chance, or whether Victoria will remain mainly a museum of past mining achievements, remains to be seen.

I am indebted to Chris Davey, former Lecturer in mining at RMIT, Gilbert Ralph, MBE, and John Reynolds of Victorian Chamber of Mines for their invaluable assistance in the preparation of this paper, which was presented in Ballarat on 10 September 1990.

Editor's Note: Changes to the Victorian Mines Act in 1991 partly remedied a number of the problems alluded to in the latter part of this address.

Mayor of Stawell, Don Webb (left) with Western Mining Corporation's Director of Operations, Keith Parry, at the official opening of the Stawell Joint Venture, 1981. Behind them is the banner of the Stawell Amalgamated Miners' Association, dated 1882.

Putting Victoria's Brown Coal to Work

In 1824, ten years before Melbourne was founded, the explorers Hume and Hovell found black coal on their way through what was to become the Port Phillip district of New South Wales and later the separate Colony of Victoria. The infant colony of Victoria in 1852 offered a reward for the discovery of a workable coalfield and in 1857 the brown coal field at Lal Lal, 100 kilometres west of Melbourne, was discovered and later developed for ironmaking. Other discoveries followed and by 1876 over 30 occurrences had been recorded, including those at Altona, Newport and Morwell, each of which was operating on a small scale in the 1890s. At that time, however, much of Victoria's energy needs were met by black coal shipped from New South Wales.

Black Coal in Victoria

In the 1890s the term Marvellous Melbourne was tempered by depression. In an attempt to become more self-sufficient, Victoria began to develop its small black coal deposits in South Gippsland. Strikes on the New South Wales coalfields in 1909 led the Victorian Government to join with private enterprise to establish the State Coal Mine at Wonthaggi.

More strikes in New South Wales in the 1930s prompted further expansion at Wonthaggi to 600,000 tons a year. At the mine's peak, over 1,700 men worked there. After the Second World War, when the Victorian Railways introduced diesel-powered locomotives, the demand for Wonthaggi coal fell and the State Mine closed in 1968.

Historical examples of black coal mining in Gippsland have been preserved. Near Wonthaggi the East Area mine has been partially reopened within the State Coal Mine Historic Reserve and at nearby Korumburra, Coal Creek has developed as a tourist venture.

Ships tied up in Melbourne during the Maritime Strike, 1890. Constant problems with the shipping of coal from New South Wales led the Victorian Government to look to local coal resources.

Development of the Brown Coal Deposits

Australia's great Maritime Strike in 1890 disrupted coal supplies from Newcastle and prompted the 1891 Royal Commission into the availability and use of Victoria's coal resources. James Stirling, Assistant Government Geologist, told the Commission that he believed the brown coal deposits in the Latrobe Valley were the largest in the world. The Government Analyst, J. Cosmo Newbery, was sent to Germany, where brown coal technology, including briquetting, was well advanced.

A great deal has been written about Victoria's brown coal deposits. In Volume One of the Transactions of The Institute for 1893 there is a paper entitled 'Notes on Victoria's Brown Coal Beds' by James Stirling. By that time several companies had open-cut mines in the Latrobe Valley, the largest being the Great Morwell Coal Mining Company, which began

Hume and Hovell expedition on its way south. The explorers found signs of coal in what was to become the Port Phillip District of New South Wales and later the separate Colony of Victoria.

operating in 1889. In 1894 the Austral Otis Company built a briquetting plant, based on the Neinburger plant in Germany, at the Great Morwell open-cut.

These developments were short-lived and brown coal production declined by the turn of the century, despite advances in electricity generation and increasing usage in Melbourne which had its first electric tramway in 1889 and electric street lighting in 1894. The better quality New South Wales coals were used in the numerous generating plants scattered around Melbourne. The Latrobe Valley deposits needed more capital to develop them and awaited further advances in electricity transmission before they could be worked profitably.

For the next 20 years interest in brown coal grew steadily. At the same time the small Victorian black coal industry, with the same union as the New South Wales fields, showed the same propensity to strike. During the First World War the strike situation in New South Wales and in Wonthaggi brought about Government action that led to the formation of the State Electricity Commission of Victoria (SECV).

Sir John Monash, GCMG, KCB, one of Australia's best known soldiers, was first Chairman and General Manager of the State Electricity Commission of Victoria, from its formation in 1921 until his death in 1931.

G. C. Drucker's excellent account of the 'History of Brown Coal Developments' in Victoria in The Institute Monograph 10 records that:

> Two events at this stage were responsible for the development of Victoria's brown coal and the realisation of the immense size and potential of the Latrobe Valley deposits. Firstly, there were the disruptions to supplies of black coal imported from New South Wales caused by World War I. Then there was another equally as important event, the appointment of Dr Hyman Herman as Director of Geological Survey. Dr Herman was more responsible than anyone for making the dream that Victoria could get electric power from brown coal a reality.

Dr Herman's report led to the appointment of three Commissioners in March 1919. The Commissioners produced a report which Sir Arthur Robinson, Minister for Public Works, described as 'one of the most valuable documents ever issued in Victoria'. It recommended an open-cut mine at Yallourn, a nearby power house and transmission lines to Melbourne.

The Commissioners engaged General Sir John Monash, GCMG, KCB, who already had a distinguished engineering career in Victoria and was still in England organising the return of Australian forces, to investigate German brown coal mining methods. Monash sent the Commander of the Tunnelling Company, Major Noel Mulligan, (who was a certified mine manager), with a party to visit brown coal operations in the Rhine Valley. By various 'cloak and dagger' methods he eventually obtained samples, photographs, plans, and diagrams that were important to future development in the Latrobe Valley.

The State Electricity Commission of Victoria was formed in January 1921 with Sir John as Chairman and General Manager. He moved with characteristic speed and efficiency and by June 1924 electricity was being generated at a 50,000 kilowatt station at Yallourn and transmitted to Melbourne at 132,000 volts, the highest in the British Empire at the time.

The history of frequent strikes by miners prompted the newly-formed SECV to exclude the miners' union from its brown coal operations. To ensure this, the words mine, miner and mining were virtually outlawed in all the Commission's documents and language. An interesting side effect was that the operations did not come under the State Mining Act. Members of The Australasian Institute of Mining and Metallurgy were affected, as 'mining engineers' were not employed by the Commission until the time of the Morwell project 30 years later.

The first sod is turned at the site of Yallourn power station, February 1921. Sir John Monash, GCMG, KCB, is second from the right.

Today the SECV is recognised as a world leader in the production of electricity from brown coal. Victoria's electricity production is at the lower end of world costs. The SECV is a net exporter of energy over the interconnecting grid that joins New South Wales and the Snowy Mountain systems with Victoria. In 1990 the Latrobe Valley region had over 60,000 people and a great majority of them derived their livelihood directly and indirectly from the Yallourn, Morwell and Loy Yang fuel and power complexes.

Resources of land, minerals and water do not in themselves produce wealth. It is the initiative, perseverance and skill of the people involved that result in the creation of wealth. An undertaking as large and as complex as this, half a world away and separated by language and wars from the only other sources of similar technology, has required leadership and skills of the highest order. The SECV has been led by very able people in its history and has been particularly fortunate that in critical periods it had leaders such as Dr Hyman Herman, Sir John Monash, GCMG, KCB, and Sir Willis Connolly, CBE.

Head of a black coal mine, Victoria.

Dr Hyman Herman is generally regarded as 'father of Victoria's Brown Coal Industry'. He studied Engineering at the University of Melbourne, then joined the Geological Survey of Victoria in 1895 as a junior field assistant, to become Acting Government Geologist by 1899. He later became Assistant General Manager of the Mount Bischoff Mine in Tasmania, but returned to Victoria as Director of the Geological Survey.

The Brown Coal Advisory Committee, formed in 1917 with Herman as Chairman, recommended the setting up of the State Electricity Commission of Victoria, based on the Latrobe Valley brown coal. When this was done, in 1921, Herman became Engineer in Charge of Briquetting and Research, a position he held until his retirement, in 1940. He was responsible for a great many innovative developments and technical studies to do with the winning the coal, briquetting it and using it in power generation.

Dr Herman was President of The Institute in 1914 and received The Institute Medal in 1938. He was made an Honorary Member in 1949 after 50 years as a Councillor. On his retirement in 1940 he wrote *Brown Coal*, a comprehensive volume on its technical and industrial aspects published in 1952. He died in 1962 aged 86, having been a member of The Institute for 65 years.

Sir John had already achieved distinction as an engineer before he was appointed Chairman and General Manager of the State Electricity Commission of Victoria in January 1921. He was born in Melbourne and after an education at the University of Melbourne disrupted by the need to earn a living was awarded a Bachelor of Engineering degree in 1891 and Bachelor of Arts and Bachelor of Law degree in 1893. He became successful as a builder of concrete bridges and other structures, using the Monier patent reinforcing system.

While at the University he had interested himself in the University Regiment and later the Militia in which he reached the rank of Colonel before the outbreak of the First World War. He was at Gallipoli and went

The start of mining brown coal from Yallourn North open-cut. The first electricity from the Latrobe Valley reached Melbourne in 1924.

on to numerous victories in Europe as Commander of the Australian Corps. He was among Australia's greatest and best known soldiers.

As Chairman and General Manager of the SECV until his death in 1931, Sir John built up one of the most successful state instrumentalities in the nation, despite numerous technical difficulties and political interference.

Sir Willis Connolly, CBE, graduated in Electrical Engineering from the University of Melbourne in November 1921 and two days later took a wages job at Yallourn. He soon became a junior Electrical Engineer in the Electricity Supply Branch, then in 1927 moved to the Briquetting Branch were he came in direct contact with Dr Herman. Connolly progressed steadily through the Sales and Electricity Supply Branches to become Assistant General Manager in 1950, General Manager and Chairman in 1956. He retired in 1971, with 50 years of service to Victoria behind him.

Sir Willis was awarded Honorary Membership of The Institute in 1970. Since 1986 The Institute, in association with the Barbarians, a group of coal mining engineers from the Latrobe Valley named after Saint Barbara, patron saint of German miners, has annually awarded the Sir Willis Connolly Memorial Medal.

The Latrobe Valley of the 1970s had the stamp of Sir Willis's leadership marked on it, as did the power industry of Victoria.

In the early 1950s, the Victorian Government had decided to expand Yallourn and develop a new project at Morwell to mine coal for four briquette factories. The new mine and factories would provide fuel for the State and raw material for a huge gas plant, making Victoria independent of outside coal sources. Within two years everything had changed. The severe recession in the aftermath of the Korean War and steep increases in wage rates, added to limited local technical knowledge and ability, brought the Commission almost to bankruptcy and led to the multi-million pound fuel and gas project being abandoned. The State Government had already taken over the Metropolitan Gas Company and created a new state-owned body, the Gas and Fuel Corporation of Victoria, in anticipation of the proposed developments.

Sir Willis Connolly, CBE, began work with the State Electricity Commission in 1921, two days after he graduated in Electrical Engineering from the University of Melbourne. He progressed steadily to become General Manager and Chairman in 1956. He retired in 1971 with over 50 years of service to Victoria behind him.

Sir Willis was well equipped to understand the situation as he had been through two world wars and a depression. By 1956 the economic scene had changed back to one of high growth in demand for power and energy in all its forms. The stop-go years of the Morwell briquetting and gasification projects gave way to predictions of power shortages and then the complications of the 1960s as the Bass Strait petroleum and natural gas discoveries made redundant much of the massive Morwell gas and fuel project concept. In the midst of all this, recognition in the late 1950s of the high sodium and sulphur content of the Morwell brown coal added to the problems.

Amongst the most significant decisions during this time was the introduction of the turnkey contract system for major projects. This had a profound impact on industry and life in the valley. The Morwell and Yallourn mining, power generation and transmission projects and the Gas and Fuel Corporation Lurgi gas plant were undertaken by contractors.

At the same time the Commission and the Government encouraged the getting together of major international plant manufacturers and Australian interests to make equipment in Australia. The policy was not always popular but was highly successful and beneficial to the Australian economy. The SECV encouraged local engineering companies to take on new work and develop new technologies, one of which in particular led to Australia becoming a world leader in conveying-equipment design and manufacture in the 1960s.

Amongst the heritage left by Sir Willis was the use of retained surpluses rather than borrowings as the major source of capital funds by state authorities, financing by savings rather than debt. This was a lesson the country was re-learning in the early 1990s.

The availability of Latrobe Valley electricity, together with the development of the Anglesea coalfield, brought about the siting of the Alcoa aluminium smelter at Point Henry in the 1960s and the building of a second smelter at Portland in the 1980s. Such industries in turn encouraged service and tertiary industries, to say nothing of the infrastructure. This flow-on of economic benefits starts with the production of energy from brown coal.

Liquefaction of Brown Coal

Attempts to produce synthetic liquid fuels from coal economically have challenged technologists since the first such process was proven over 70 years ago by the German scientist Berguis. Germany built several plants to reduce its dependence on imported petroleum. In 1935 Dr Herman initiated a programme of research, at the Fuel Research Centre in England, into the viability of converting Latrobe Valley coal into liquid fuel. Results of this work were kept secret on completion in 1939 for fear that it might be of value to Germany, which who had embarked on a similar development in Czechoslovakia to provide fuel for the Second World War.

It took the OPEC oil crisis of 1973 to re-awaken interest in the possibility of liquefaction. A joint Australian-West German re-evaluation of the project was undertaken but did not lead to development. The next step came in 1980, when the Japanese, Australian and Victorian Governments agreed to co-operate in research for brown coal liquefaction. By 1988 a pilot plant, capable of producing about 50 barrels of liquid fuel per day, had been built at Morwell at a cost of $500,000,000. It was the largest such plant in the world for brown coal liquefaction and, after several years of intensive research aimed at optimising the process, work was discontinued in the early 1990s, for while it was technically feasible it was not economically viable at the time.

Bucket wheel coal excavator at Morwell in the 1970s.

Yallourn W power station in the Latrobe Valley has a capacity of 1,450 MW. The first unit began operating in 1973.

Research into liquefaction of brown coal has been carried out by Brown Coal Liquefaction (Victoria) Pty Ltd since 1980.

Value of Brown Coal

While it is not possible to put a meaningful value on its production of coal and electricity, and the potential for liquefaction, there is no doubt that the Latrobe Valley has been and will continue to be of immense benefit to Victoria. Of the estimated 52 billion tonnes of brown coal which is readily recoverable there less than two per cent has been used. The deposits are a great national asset.

Brown coal is, however, not the only valuable energy source of the region.

Bass Strait Oil and Gas

In 1924, small showings of oil and gas were obtained in a borehole drilled near Lakes Entrance in Gippsland. More than 50 bores were drilled and some very heavy oil obtained. It was no bonanza, but during the Second World War the Victorian Railways used some oil from these bores.

Post-war development in Australia and the almost complete dependence on imported oil prompted renewed efforts in exploration for petroleum throughout Australia. Known and potential areas such as Roma, the north-west of Western Australia and southern Victoria were natural targets.

In 1954 BHP's subsidiary, Australian Iron and Steel Ltd, was seeking oil and gas in the southern part of the Sydney Basin, without success. Five years later, Sir Ian McLennan, KCMG, KBE, then BHP's Chief General Manager, decided it was time to seek advice from the United States. What happened is now an important part of our mineral heritage. Alan Trengove in his book *Discovery* records how John Norgard, a BHP General Manager, visited Lewis Weeks, a retired Chief Geologist from Standard Oil Company of New Jersey. Weeks had never been in Australia, but for 40 years he had studied the world's sedimentary basis and knew something of the geology of Australia. Pointing to a map of Australia he said:

> You are in the wrong area. You've got to get your feet wet. The oil-bearing rocks in Australia are the younger rocks, down in the south, and out in the water.

By this time the technology needed to search for and recover oil and gas from offshore was available. BHP engaged Lewis Weeks as a consultant and applied for leases in Bass Strait.

Airborne surveys and marine seismic surveys were completed by December 1963. BHP and Esso began their successful association in mid-1964 and the first offshore well discovered the Barracouta gas field in February 1965. Gas and oil were found in the Marlin field in 1966, oil at Halibut in 1967 and in the giant Kingfish field in 1967.

The first cargo of Bass Strait crude oil was loaded from Long Island Point on 23 March 1970, at a price of $2.10 per barrel and the fractionation plant was opened in the same year. In 1985 Bass Strait production

Drill rig Ocean Digger *in Bass Strait, about 1970.*

reached a one day peak of 537,000 barrels. The price peaked at $44.97 per barrel, but fell rapidly during the next year.

In the 20 years after the first shipment, 2.6 billion barrels of oil were produced from the field, enough to meet 57 per cent of Australia's petroleum needs. In addition, 325,000,000 barrels of LPG and 2,800,000 tonnes of ethane were recovered. These resources saved the Australian economy more than $109 billion in 1990 dollars, and payments to Governments through excise and royalties amounted to more than $72 billion.

Future Outlook

Most Victorians who use electricity, natural gas and petroleum have little concept of where they come from and how they are produced. They merely flick a switch, turn on a tap or pull up at a bowser. Little do they care about how it all happens, except when it does not happen.

It is not uncharacteristic of people to be like this and that is why I believe it important that we should all speak up whenever we can about the contribution the resources industries make to the community. We can all be very proud of our part in this. Our individual voices, sounded often enough, and provided there are enough of them, will add up to a powerful chorus.

I am indebted to Doug Stewart, former Chief Engineer – Fuel, SECV, and Gilbert Ralph, MBE, for their invaluable assistance in the preparation of this paper, which was presented in Morwell on 9 April 1990.

Part of the oil and gas pipeline system on the Esso-BHP Barracouta platform, Bass Strait, 1970s.

Tasmania

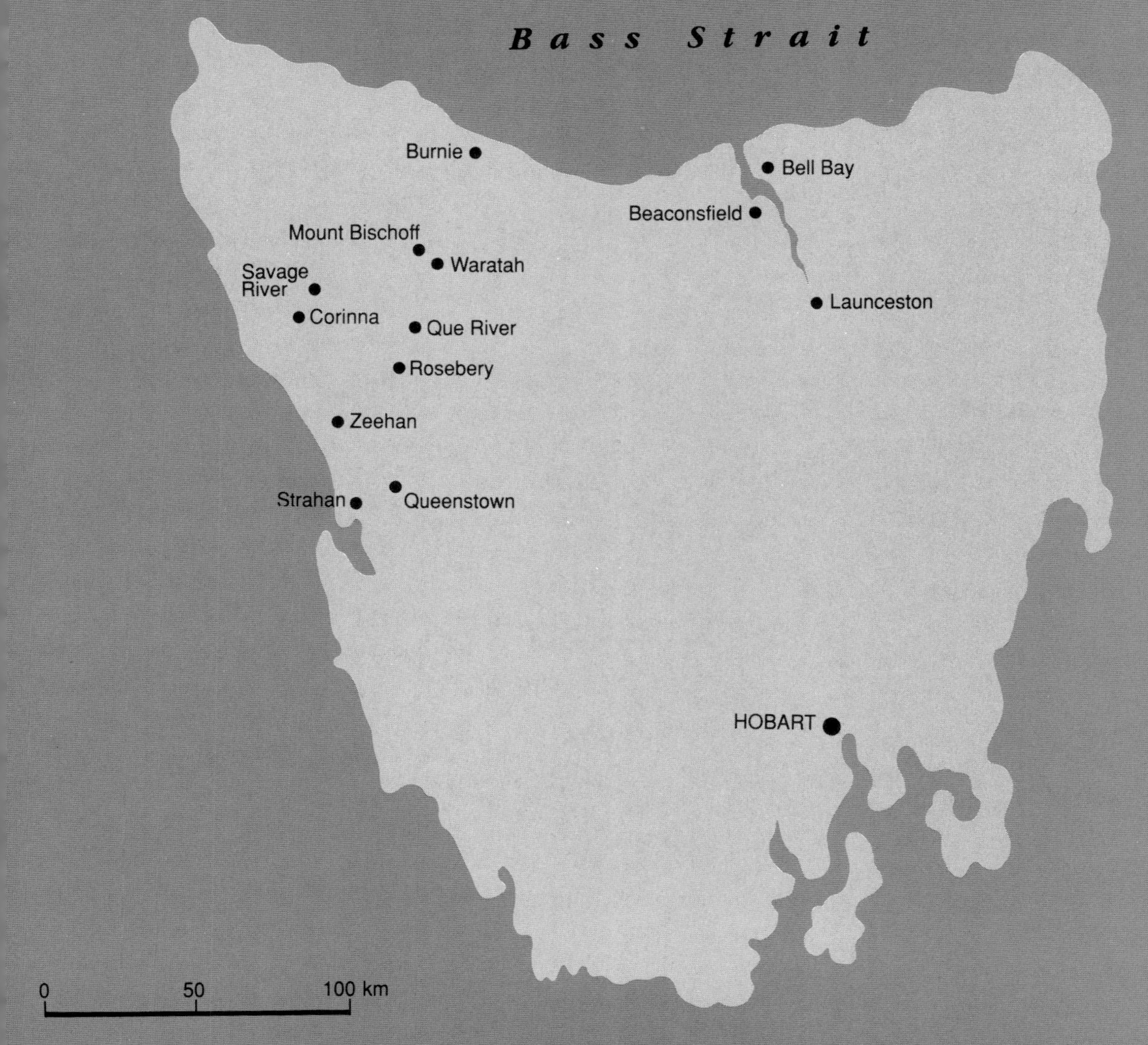
King Island
Flinders Island
Bass Strait
Burnie
Bell Bay
Beaconsfield
Mount Bischoff
Waratah
Savage River
Corinna
Que River
Launceston
Rosebery
Zeehan
Strahan
Queenstown
HOBART
0
50
100 km

People Make It Happen

Edward Hammond Hargraves, officially credited with being the first person to find gold in Australia, was honoured in this 1951 stamp.

Most mining activity takes place a long way from the main urban centres of population. Miners on the road to mountain diggings.

People make a very appropriate theme for a talk to a Branch of The Australasian Institute of Mining and Metallurgy, for people are what The Institute is all about. It is the organisation of the technological professionals in the industry, as distinct from the Australian Mining Industry Council and the Chambers of Mines, which are organisations of the companies; AMIC on an Australia-wide basis and the Chambers in individual States.

The technological professionals are complemented by people in other professions and the non-professionals, all of whom are indispensable and make vital contributions. I concentrate on the technological professionals in this talk, only because The Institute is their organisation and not because I overlook the value of the others in the industry.

The Institute

The Australasian Institute of Mining and Metallurgy was founded in 1893 in Adelaide, just over 50 years after the discovery of Australia's first metal mine at Glen Osmond in South Australia in 1840.

The aims of The Institute, as expressed in the Royal Charter of 1955 are, in an abbreviated form, to:

- Promote and advance the science and profession of ... mining and metallurgy in all its branches and to facilitate the exchange of information and ideas;
- Raise the character and status and advance the interests of the profession of ... mining and metallurgy and those engaged therein;
- Increase the confidence of the community in the employment of recognised engineers;
- Promote honourable practice;
- Collect and circulate statistics and other information;
- Provide for the delivery and holding of lectures, exhibitions, public meetings, classes and conferences to advance education in engineering;
- Encourage the study of ... mining and metallurgy and to improve and elevate the general and technical knowledge of persons employed or intending to engage in the profession;
- Communicate to members information on all matters affecting the profession;
- Encourage the discovery of and to ... make known the ... merits and processes ... capable of being used by persons engaged in the profession;
- Promote improvements in the law (relating to the industry); and
- Promote and safeguard the interests of the profession.

In its now nearly 100 years of existence, The Institute has been a highly respected body, fostering technical progress and excellence, and nurturing professional and ethical standards, professional education, and the professional standing of its members.

It has been less active in participating in public policy formulation affecting the industry, but this is changing in recognition of the changing political and community environment. Policies are becoming far too important to be left to the politicians.

In 1990, the 38 Branches had a total of 7,840 members from a wide spectrum of the technical professions in the industry: mining engineers, metallurgists, geoscientists, chemical engineers, engineers and other related professions.

The People

The mineral industry shares with the pastoral and much of the rural industry the fact that most of its activities take place a long way from the main urban population centres in the coastal areas. One result of this is that the views held by many Australians about the industry are not well informed, to say the least.

It also means that the people who work in the industry, and their families, spend much of their lives in these remote areas, away from the large cities. Most city-dwellers today – and this means most Australians

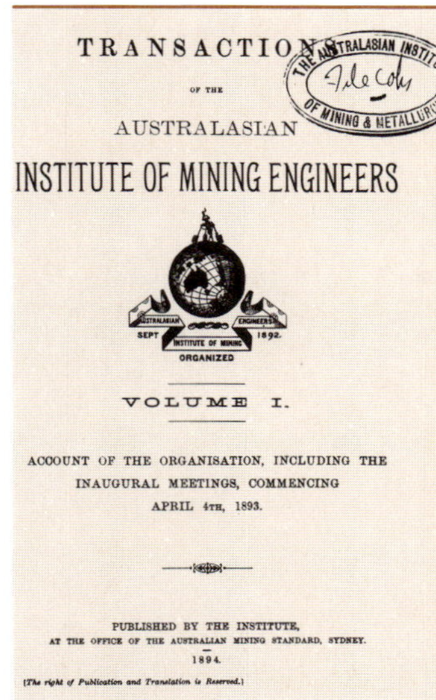
TRANSACTIONS

OF THE

AUSTRALASIAN

INSTITUTE OF MINING ENGINEERS

VOLUME I.

ACCOUNT OF THE ORGANISATION, INCLUDING THE INAUGURAL MEETINGS, COMMENCING APRIL 4TH, 1893.

PUBLISHED BY THE INSTITUTE, AT THE OFFICE OF THE AUSTRALIAN MINING STANDARD, SYDNEY.

1894.

(The right of Publication and Translation is Reserved.)

Title page from Volume 1 of Transactions of the Australasian Institute of Mining Engineers, the body which became The Australasian Institute of Mining & Metallurgy. The Institute has always been a 'people' organisation.

Shift bosses at the No. 2 shaft, Broken Hill South mine.

– would regard this as unattractive, if not unacceptable. Many people in the cities even regard having to take a job in a distant suburb as unreasonable. A long period of prosperity and the growth of an entitlement mentality have distorted their perceptions of the realities of life.

Gustav Thureau, Government Geologist in Tasmania, 1882-89.

Most people in remote locations enjoy what they are doing and the lifestyle. Many like the freedom of the open spaces, the often harsh but, to them, attractive environment of the desert or the bush, the informal social atmosphere and the absence of city noise and traffic jams. Family members sometimes find it hard to do without the city lights, but the improvements in modern communications and travel and the coming of satellite television have greatly eased this problem.

The small number of Australians in the mineral industry form a rare group of people, in my view not only more down-to-earth, enterprising and self-confident than many of their urban cousins, but also rendering absolutely essential, and mostly unrecognised, service to Australia. The Commonwealth Statistician in 1989 listed just over 107,000 people as working in Mining. They and their families produce the products which earn more than 40 per cent of the export income that underpins the living standards of 17 million Australians.

As this talk is being given in Tasmania, I think it is appropriate to quote a passage describing the people who pioneered the industry in this State last century:

> These miners and prospectors are a fine class of men, intelligent and hardworking and the very perfection of hospitality. The bacon and damper and any other delicacies at their command are offered you with a heartiness that touches you, when you think that they have carried them on their own backs for miles through the scrub. Few people know how much the country owes to these manly and independent fellows and no one who hasn't worked his way along a rough prospector's track for a day or two at least, with house, clothing and food on his back can realise the hard conditions under which they do their work of pioneering.

Copper miners, South Australia, drilling blast holes with a compressed-air powered drill, circa 1916.

Conditions are of course vastly improved today, but the spirit remains. The industry is led by the professionals in it, who have always needed and will continue to need high technical ability. Another characteristic of this industry is that the people in the senior professional positions also need high management skills, particularly the ability to lead others. The remote nature and small numbers of people in most mineral operations and communities demand a much greater leadership role from them than in similar jobs in a capital city.

The Women In the Industry

Mining has traditionally been seen as a man's occupation, although, as wives and family members, women have shared the hardships and the excitement, the disappointments and the triumphs since the earliest days. Their contribution to the industry cannot be overstated.

Early mining fields did not encouarge family life, but it was not entirely absent. This family on the Croydon field in Queensland seem to be setting off on an outing, possibly to visit friends, about 1890.

In 1986 the Perth Branch of the Women's Auxiliary of The Institute produced a diary as a tribute to the 'Mining Women' who helped develop and establish our mining towns. The Diary contains many delightful stories recalled by those who took part in the developments, sometimes as children.

In recent times, in keeping with changes in general community attitudes, more and more women have singly entered the industry in both professional and non-professional roles. Amongst The Institute's 7,840 members in 1990 there were more than 200 women, in disciplines ranging from geoscientists to mining engineers and metallurgists. The numbers while still small, are growing.

Whatever prejudice there may have been against women in the mineral industry, I am glad to say it no longer exists. The Institute itself has a proud record in this in that the late Beryl Jacka, AM, MBE, as Secretary, headed the administration of The Institute for 30 years and our present Chief Executive Officer is Judy Webber.

Beryl Jacka, AM, MBE, who as Secretary, headed the administration of The Australasian Institute of Mining & Metallurgy for 30 years.

The Stories

A prominent place in Australia's mineral heritage is occupied by the enormous fund of stories about the industry, the successes and failures, the marvellous discoveries and the unsuccessful endeavours, those who struck it rich and those who did not. The stories are more about people than events and are rich in the wonderful humour so characteristic of people in industry.

We have a growing literature of the history of Australian mining, but much more remains to be recorded. May I suggest that members of the

Workers at Butterfly Shaft, Norseman Gold Mines N.L. circa 1950.

AusIMM, most of all those in positions of authority in the industry, have an obligation to encourage and assist the recording and preservation of our rich past. It is by knowing, respecting and drawing inspiration from the past that we can cope with the future in a way in which our successors can be in due course proud of us. It is pleasing that in addition to writings, there is also a small but growing record of Australian mining in paintings, drawings and etchings. Music so far is poorly represented.

It is impossible to do more than give just a few snippets of the history and the humour in the time available in a short address. Needless to say, the samples are picked at random and are not comprehensive or representative.

The cradle of Australian metal mining was copper mining in South Australia, where the discoveries of the 1840s blossomed into an industry producing for a time 10 per cent of the world's copper requirements. The miners were mainly Cornish, and the area around Moonta was referred to as Australia's Little Cornwall.

C. J. Dennis, best known as the author of *The Sentimental Bloke*, suggested to Oswald Pryor (who began working in the mines at Moonta at the age of 13 and eventually became a manager), that he should record the Moonta Cousin Jacks and Jennies in his cartoons. Pryor's cartoons, many published in the *Bulletin*, are a marvellous record of some of our mining humour.

Miners at Emperor gold mine, Fiji, 1966.

When the Moonta mines closed, many of the miners and their families moved to Broken Hill. Some walked, pushing their belongings in front of them in wheelbarrows. This journey has been in recent years re-created with a 'barrow race'.

In more recent times, one of the mining executives in Broken Hill was Struan Anderson. As underground manager he was on a visit underground when suddenly a man came tumbling down an ore pass from the level above and fell in a heap at Struan's feet.

'Good God man, where did you come from?'

The Cornish miner struggled to his feet:

'I cum from Moonta, Sir!'

Some years later it is said that Sir James Foots happened to come across four card players who had overstayed their lunch break in the underground crib room. Euchre, the favourite miners' card game in Broken Hill, was in progress. One player had a good hand.

'I'll go alone', he said.

Sir James heard the comment as he came into the crib room.

'No you won't. You all go together'.

The major mineral discoveries in Tasmania last century, starting with tin at Mount Bischoff in 1871, made it the second base-metal province found in Australia. There is a good record of the times in several excellent books.

Miners going underground at Kalgoorlie, 1930s.

The story of Mount Morgan in Queensland in the same period would fill many evenings. Again, it is fortunately recorded in several books as is the story of Broken Hill and, to a lesser extent, Kalgoorlie.

Of the many thousands of prospectors, some were successful in finding ore, such as Dutton of Kapunda, 'Philosopher' Smith of Mount Bischoff, Hargraves of Bathurst, Miles of Mount Isa, Hannan of Kalgoorlie, Rasp of Broken Hill and so on. For every success there were thousands of failures. Even successful prospectors rarely made personal fortunes. Their legacy was an industry which continues to serve Australia well today.

There are more stories waiting to be told about the more recent finds and developments: about iron ore, manganese, aluminium, oil and gas, nickel, copper and uranium, coal, mineral sands and so on. It is essential that they be told. It bears repeating that it is to a large extent the responsibility of members of The Institute to see that this happens.

In 1935 the AusIMM established The Institute Medal, which has been awarded on 50 occasions to members considered to have rendered service of exceptional merit to the Australian minerals industry. The first was Guillaume Delprat of Broken Hill and Newcastle steel industry fame. The list since then contains a veritable Who's Who of illustrious names in the industry. It must be remembered that only those who made their main contribution since the Medal was established in 1935 are included.

The Future

Sir Herbert Gepp was one of the great leaders of the industry in Tasmania. In his Presidential Address to The Institute, in Kalgoorlie in 1933, he said:

Electrical apprentice at the Trades Training Centre at Mount Isa Mines.

> As representatives of one of the principal primary and secondary industries in Australia, the members of the Institute have their duty to perform as citizens – to co-operate fully in contributing to the well-being of the nation.
>
> By building up a sound and logical system of mutual understanding between all engaged in the industry, by taking thought for tomorrow in regard to the training and encouragement of the younger men who will be later the leaders of the industry, by continual research in all directions, and by making their united voice heard in the councils of the nation, the Institute through its members, collectively and individually, can show themselves worthy of their privileges and their inheritance.

Many people have found employment in industries peripheral to the mining and mineral industry. This sharebroker's advertisement is from an 1888 publication.

> The duty of our industry is to aid the sound development of our heritage in which lies the future safety of our children and children's children.

There is still nothing to add to this masterful summation. I would only like to expand on some of the points.

The training and encouragement of younger men (and today Sir Herbert would add women) has never been more important than it is now. We have first-class people in the industry today, but the at-times literally defamatory attacks against the industry by small but noisy groups threaten that. They create the danger of other occupations coming to be seen as offering a higher standing in the community, mainly by those who live in the big cities and do not know the industry. There is the danger that the minerals industry may not attract enough of the best young people in the future.

The Institute is doing something about this. The AusIMM Education Endowment Fund, established in 1989, is focussed on attracting high calibre students to study the disciplines relevant to the industry and on establishing and maintaining contact between the industry and the students during their studies.

The effort to counter the negative image created by the groups opposing the industry will be stepped up. We plan to increase the membership by encouraging the numerous professionals qualified but not yet members to join. A substantial increase in membership will help to enhance the standing of The Institute while in turn making it easier to represent the members.

I am grateful to John Dew and Gilbert Ralph, MBE, for their assistance in the preparation of this paper, which was presented in Hobart on 15 February 1990.

Tasmania: An Important Mineral Province

Tasmania is a very important mineral province, with 20 major ore deposits discovered in the last 120 years. In terms of timing, Tasmania was the second major base-metal district discovered in Australia, after the South Australian copper finds in the 1840s.

The first mineral found in Tasmania was coal, on the south coast in 1793, followed by more coal elsewhere. It was 1834, however, before any of the coal deposits were developed. Although gold was found near Launceston in 1847, it was not until after the Victorian rushes that gold mining began at any scale. A major gold mine operated at Beaconsfield, across the river from Bell Bay, towards the end of the 19th century and in the early part of the 20th century.

Remains of the Beaconsfield gold mine, on the Tamar River, northern Tasmania. Gold was first found there in 1847, with a major strike in 1877. The mine was down to 450 metres when pumping out the inflowing water became too costly. It closed in 1913.

The key mineral discovery in Tasmania was tin at Mount Bischoff in the north-west in 1871. An operating company was formed in 1873 with 12,000 shares called up to £1 each. The mine was a great success, becoming for a time the largest tin mine in the world. Dividends paid totalled over £200 a share, and shares traded on the Stock Exchange for years at £80. James 'Philosopher' Smith, who had discovered Mount Bischoff, sold the shares allotted to him before the first dividend was paid. He also received £1,500, a silver salver and an illuminated address. He died a poor man in 1897.

The Mount Bischoff discovery stimulated the search for minerals, not only along the West Coast but also in the north-west of the State. More major discoveries followed before the end of the 19th century, including the famous orebodies of gold-copper at Mount Lyell, silver-lead at Zeehan, lead-zinc at Rosebery, and tin at Renison Bell. Renison was the world's largest underground tin mine in 1990.

We have excellent accounts of mineral discoveries and mining activities in Tasmania, including Professor Geoffrey Blainey's *The Peaks of Lyell* and *The Rush That Never Ended*. In this brief address I can do no more than hint at the colourful stories, the romance, the hardships and the achievements.

Periods of feverish activity on the Stock Exchanges were a feature of this period, when there were Stock Exchanges at Hobart, Launceston, Zeehan and Queenstown. It is reported that on one occasion in 1879 the Hobart Stock Exchange stayed open all night, a feat probably not repeated since. Much of the speculation, then as now, was based on rumours and the herd instinct.

'Philosopher' Smith, who discovered tin at Mount Bischoff, Tasmania.

Wild as the speculative booms of the late 1800s may have been, they did have the useful effect of making capital available for the real mineral developments.

Tasmania's mineral discoveries and developments this century have been less spectacular but equally significant. Of the three major ore

Crushing batteries at Mount Bischoff. The tin deposits James ('Philosopher') Smith found here in 1871 led to the development of the world's richest tin mine.

discoveries in Australia in the first 25 years of this century, two were in Tasmania: the King Island scheelite deposit and the Aberfoyle tin-wolfram deposit. The third was Mount Isa in Queensland. Production at King Island began during the First World War, expanded during and after the Second World War and recently closed.

The second period of significant ore finds in Tasmania this century has been since 1965, including discoveries such as tungsten at Bold Head and Kara, lead-zinc-silver-gold at Que River, tin at Severn, lead-zinc-silver-gold at Hellyer, gold at Henty and silica at Corinna. Substantial additional ore reserves were proven at Savage River (iron ore), Cleveland (tin), Renison (tin), and Mount Lyell (copper-gold). The existence of iron ore at Savage River has been known since 1877, but it was not until the demand for iron ore by the burgeoning Japanese steel industry that development began in 1965.

Working on the face at Mount Bischoff from an early etching.

Dr Ross Large of the University of Tasmania recently estimated the in-ground value of the discovered mineral resources in western Tasmania at $31 billion. About 60 per cent of the total ore discovered to date has been mined. Some 1,750 mineral occurrences are known in Tasmania. The most significant metals in terms of estimated present day value of production to date have been: zinc $4.2 billion, copper $3.4 billion, tin $2.9 billion, gold $2.0 billion, lead $0.7 billion and silver $0.6 billion.

Terrain and Climate

The island state's major mineral finds all occur in western Tasmania. Two features made mineral exploration and development in this area quite different from that in any other part of Australia: there was no surface gold (other than odd quartz reefs of little significance), and the rugged terrain and climate deterred all but the most desperate or courageous prospectors. The areas of high mineral potential are mountainous, dissected by deep valleys and ravines, thickly forested and subject to heavy rainfall. But in spite of the very difficult terrain, the riches or expected riches of the finds last century attracted the prospectors, the companies and the capital. Railways often pre-dated roads and were the main means of access and transport for people and for materials.

There is a delightful story told by Ron Midson, who worked on the West Coast 50 years ago, about the company chairman and another director, accompanied by their wives, travelling by train to visit a mine. The fireman was required to list any travellers. His report for that journey was: 'Passengers: two adults and two adultresses'.

Bringing supplies to the early mining fields was more than difficult. In 1877 a small vessel met such bad weather that it required 79 days to sail from Hobart to Macquarie Harbour. The journey through the mountain to the mines took many more days.

Mining plant at Rosebery, 1920s. The zinc content of the ore at Rosebery was a problem until 1917, when the Electrolytic Zinc Company of Australasia Limited (EZ) began refining zinc at Risdon, near Hobart.

Infrastructure and Other Industries

On the West Coast, going from Waratah south to Port Davey, nearly every settlement has its origin in the mineral industry, with the exception of a few based on fishing. The Port of Burnie was developed initially to supply Mount Bischoff and to ship the tin concentrate to the smelters in Launceston. On the West Coast, Macquarie Harbour and Trial Harbour were both established to reach and supply mining operations, and ship out their products.

Benefiting from the development of the railway lines were the timber cutters, following behind the miners to provide the needs of the mines and the railways and the ports. The opening up of the immense forests of the West Coast gave rise to forest product industries. These industries continue to be of great significance to Tasmania and to Australia.

Substantial towns were established quickly following mining developments. Zeehan, for example, grew up after the discovery of rich silver in 1884. The railway arrived in 1892 and the town had a stock exchange, newspapers, eighteen hotels, theatres, banks, churches and a School of Mines.

Another of Tasmania's major industries owes its origins to mining: pulp and paper making using indigenous species of eucalypt. In 1925 Amalgamated Zinc (de Bavay's) Ltd initiated major testing of wood from north-west Tasmania for making paper. From this emerged in 1936 at Burnie, a new enterprise, Associated Pulp and Paper Mills Ltd, making fine papers under the sponsorship of Amalgamated Zinc Ltd, North Broken Hill Ltd, Broken Hill South Ltd and Consolidated Zinc Ltd. This was followed in 1941 by the Australian Newsprint Mills project at Boyer, to satisfy some of the increasing Australian demand for newsprint. Australia is now about 70 per cent self-sufficient in paper and paperboard, and Tasmania is not only a major producer but also a significant beneficiary.

Tom McDonald, who discovered zinc-lead at Rosebery. Born in Tasmania, he prospected for gold on the mainland at such places as Oodnadatta, Katherine, Alice Springs and Broome. In 1893, he went home across Bass Strait and formed the Rosebery Prospecting Association, which took its name from the British Prime Minister. Instead of gold, he found a zinc-lead ore body.

Smelting works at Mount Lyell mine. First pegged as a gold claim, in 1884, the mine proved to be Australia's largest deposit of copper known at the time. The Mount Lyell Mining and Railway Company Limited was formed in 1893 to work it.

Mineral Processing in Tasmania

The major base metals produced in western Tasmania – copper, tin, lead and zinc – have had an impact on industrial activity throughout the nation. In Tasmania, apart from early copper smelting at Mount Lyell, the major mineral processing development related to these metals is the electrolytic zinc refinery at Risdon near Hobart. Savage River iron ore concentrate is pelletised at Port Latta.

In *Technology in Australia 1788-1988,* published in 1988 by the Australian Academy of Technological Sciences and Engineering, G. B. O'Malley describes the early work done at Mount Lyell, including the remarkable work by Robert Sticht 'culminating in the first successful operation in the world of full pyritic smelting without added fuel and with cold blast air in the furnaces'. This process resulted in the hills around Queenstown being denuded of vegetation by the sulphuric acid formed by the sulphur gases combined with the high rainfall. The unusual coloration of the rocks became a tourist attraction. During my visit to Mount Lyell some years ago some locals were becoming concerned that the vegetation was starting to grow again. Mount Lyell is also the birthplace by accident of the hydraulic mine fill technique, later perfected at Broken Hill South.

Tasmania's west coast mining was notable for the number of small railways that provided it with transport in difficult country.

A by-product of copper mining, pyrites, has also left its mark with the formation of a company well-known to generations of Australian farmers as the Wallaroo Mount Lyell Fertilizer Company, which owes its origin to both the early South Australian and Tasmanian copper mining industries. The name is passing into history because of a takeover, but the plants at Wallaroo and elsewhere in South Australia continue to operate. The availability of sulphuric acid was an important reason for establishing the titanium dioxide plant at Heybridge.

O'Malley also refers to the exceptional achievements in electrolytic zinc refining at Risdon in greatly improving the process and adapting to meet the requirements of Broken Hill's unique ore since the plant was established in 1919. Tasmanian technologists have contributed in recent years to cassiterite recovery using heavy media separation and flotation.

The first hydro-electric generating plant in Tasmania was installed in 1893 by the Mount Bischoff Tin Mining Company at Waratah. The availability of hydro-electricity was the main reason for siting the zinc refinery at Risdon, as it was for establishment of aluminium, ferro-manganese, and ferro-silicon production at Bell Bay in more recent years. The combination of high mountains and an abundant rainfall is a valuable natural resource for Tasmania.

Headframe at Mount Lyell in modern times.

There are fascinating stories about all these developments but, again, time does not allow going into detail. The one aspect which must be mentioned is that Bell Bay is the birthplace of the Australian aluminium industry.

A small alumina refinery and aluminium smelter, a joint venture by the Commonwealth and Tasmanian governments, began production at Bell Bay in 1955. The venture was seen mainly as a strategic defence measure to end Australia's complete dependence on overseas supplies of the metal and to acquire the technology and skills of aluminium production. Because bauxite was not then known to exist in commercial quantities in Australia, it was imported from Malaya and Indonesia.

From this small beginning, Australia has gone on to become by far the world's largest producer of bauxite and alumina, as well as the fourth largest producer of aluminium metal.

Significance to Tasmania

In 1990, Tasmania's mining industry generated income of $1000 million a year, 63 per cent of the State's exports and 14 per cent of its

Electric converters at the zinc refinery, Risdon, near Hobart, 1920s.

Zinc refinery, Risdon, near Hobart. The plant, set up by the Electrolytic Zinc Company of Australasia Limited (EZ), began operating in 1917, to refine zinc ores from Broken Hill, using cheap electricity from Tasmania's newly-established hydro-electricity system.

In the mining museum at Zeehan. Silver-lead discoveries in the area in 1882 led to the growth of a busy mining centre, with the period of greatest prosperity from 1893 to 1908. After that mining, and the town, declined rapidly.

gross domestic product. The Tasmanian Chamber of Mines estimate that the industry account for 7,500 direct jobs and a further 12,900 indirect jobs, equivalent to 10.8 per cent of the State's workforce.

The economic contribution of the Tasmanian mining industry is the product of the spirit, perseverance, character and achievements of the people it has attracted over the last century and a half. The amazing stories of some of those known by name – the leaders and the colourful characters – are recorded. We can be certain that there are many other impressive achievements which go unrecorded.

In the knowledge that I cannot possibly be comprehensive or do justice to the subject, let me mention a few who come to mind readily.

Most of the people I will mention were very active in Institute affairs, many as Presidents, and many were recipients of the Institute Medal.

The discovery of Mount Bischoff, which is the start of substantial mining in Tasmania, is credited to James 'Philosopher' Smith, a farmer turned prospector. He is said to have been disappointed at first that the black mineral he found was not silver.

After the discovery of Broken Hill in 1885, links were established with Tasmania. These included Bowes Kelly, several times Chairman of BHP, who was involved in setting up the Mount Lyell Mining Co. N.L. in 1892.

Early Australian mineral discoveries attracted people from overseas who made a great contribution. In Tasmania there was a strong German influence in the late 1800s. Most of them were graduates of the Clausthal Mining Academy in Germany, one of the two world eminent mining schools at the time. Many of the American mining engineers and metallurgists of the time were educated at Clausthal.

Names such as Robert Sticht, the metallurgist responsible for pyritic smelting at Mount Lyell, Dr Abt who introduced the mountain climbing railway from Macquarie Harbour to Queenstown named after him, and Heinrich Keyser, who brought the early struggling Mount Bischoff mine to profitability, belong to this group.

Dr Hyman Herman, manager of the Mount Bischoff mine around the turn of the century, went on to become the father of the Victorian brown coal industry. He was one of the first to be awarded The Institute Medal after it was introduced in 1935, was President of The Institute, and was awarded, very appropriately in Hobart, Honorary Life Membership in 1949, after having been a councillor for 50 years. He had been a member for 65 years at the time of his death, probably an all-time record.

Moving on in time, Sir Herbert Gepp, as General Manager of the then Electrolytic Zinc Company, was responsible for setting up the Risdon zinc refinery. Harry Hey, President on three occasions, worked with Sir Herbert Gepp in Risdon and became the Managing Director of EZ in 1947. A famous name in Queenstown and Rosebery is Murray, both father and son. Russell Murray succeeded Robert Sticht as General Manager at Mount Lyell and his eldest son, Hugh Murray, followed in his father's footsteps to become General Manager in 1948.

The long time Managing Director and Chairman of Western Mining Corporation, Sir Lindesay Clark, AC, KBE, CMG, MC, grew up on the West Coast, where his father Lindesay Clark was a mine manager. Sir Henry Somerset, CBE, of Associated Pulp and Paper Mills has a lifelong association with Tasmania and the mineral industry.

King Island Scheelite mine, 1949. Production of scheelite began here during the First World War and expanded during and after the Second World War.

Whyte River gold dredge during erection, 1900.

A 55-tonne capacity ore truck emerges from the main decline at Mount Lyell. The Mt Lyell Mining & Railway Co Ltd is now owned by Renison Goldfields Consolidated Ltd.

Future Outlook

Dr Ross Large points out that the total ore reserves of all the mines in Tasmania are now less than the production to date. A depletion of over 60 per cent of the total known resource has occurred. To quote Dr Large:

> Because of the increased rate of current mining, it is necessary for new mines to be discovered on a regular basis in the future if the industry is to survive at its present level. It is estimated that one major discovery every five years is needed to maintain a healthy mining industry in Tasmania. This is about the same discovery rate as that achieved in the previous 120-year history of the industry. However new mines are becoming harder to find, and therefore a significantly increased effort is required.

Whether this effort is applied depends entirely on whether the Government and the community in Tasmania recognise the vital need for the mineral industry to continue and to grow. There can be no doubt that there is more ore to be found. Attempts to restrict or prevent the exploration for minerals have been successful both in Tasmania and elsewhere in Australia. The Australian Mining Industry Council estimates that 23.5 per cent of Australia's land area is now unavailable for mineral exploration or available only under severe constraints. There are strong pressures to exclude further large land areas.

On a national scale, mining provides more than 40 per cent of Australia's export income on which the living standards of 17 million Australians depend. It is important that we, the members of The Institute, support the good work The Tasmanian Chamber of Mines is doing and help by using every opportunity to express our individual views. We have every right to be proud of our mineral heritage and of the irreplaceable contribution the industry is making to improving the lives of Australians.

I am indebted to John Dew, Doug Stewart (retired mining engineer of Melbourne), and Gilbert Ralph, MBE, for their invaluable assistance in the preparation of this paper, which was presented in Burnie on 12 February and Zeehan on 13 February 1990.

New Zealand

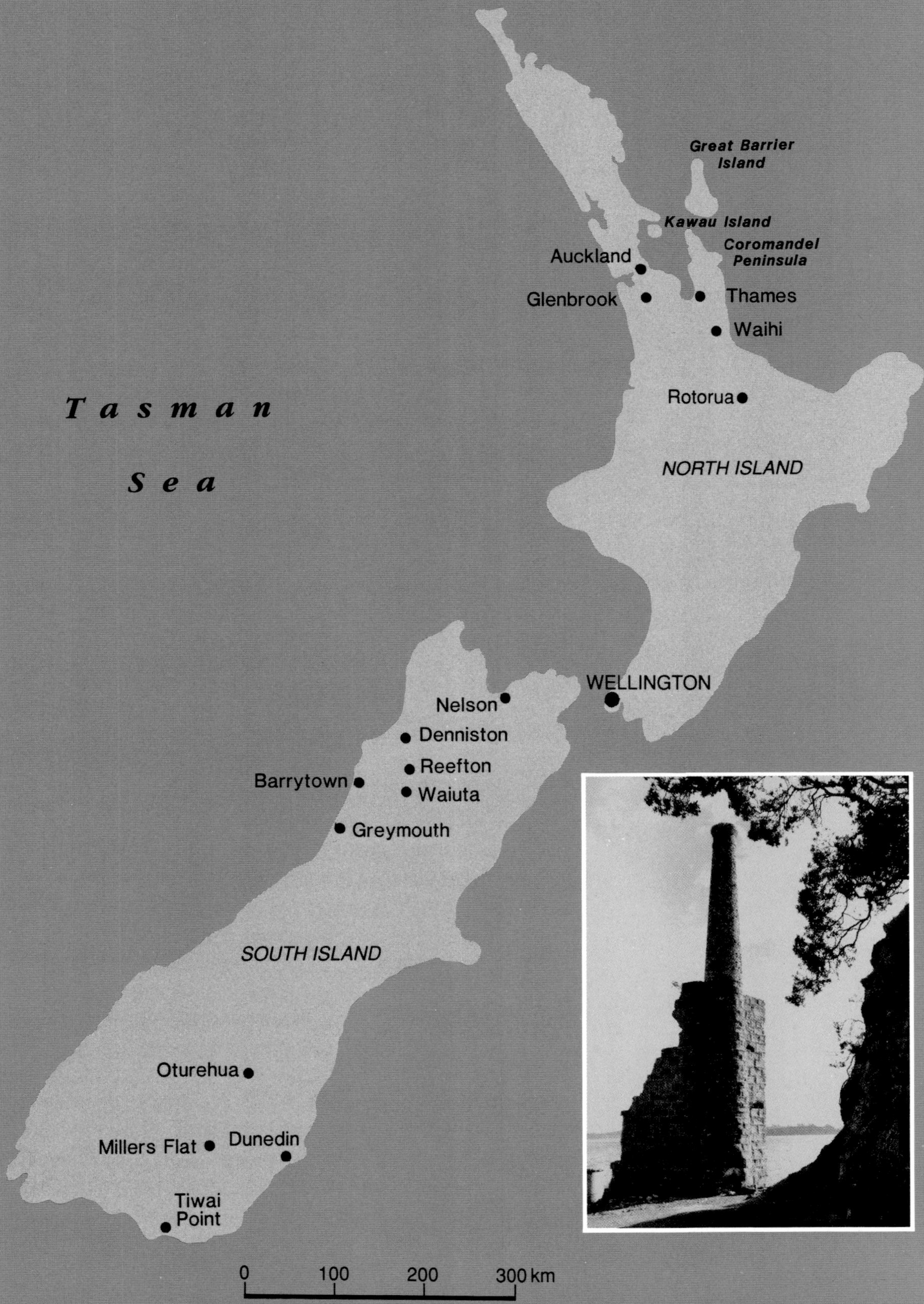
Tasman Sea
Great Barrier Island
Kawau Island
Coromandel Peninsula
Auckland
Glenbrook
Thames
Waihi
Rotorua
NORTH ISLAND
WELLINGTON
Nelson
Denniston
Reefton
Barrytown
Waiuta
Greymouth
SOUTH ISLAND
Oturehua
Millers Flat
Dunedin
Tiwai Point
0
100
200
300 km

New Zealand's Mining Heritage

Mining in all its forms has had a very significant effect on the development of New Zealand. Its history has been affected economically, technologically and socially.

The earliest mining was in the 1840s for manganese, soon to be replaced by copper on Kawau and Great Barrier Islands. Small scale copper mining took place in the Dun Mountain and Roding River area of Nelson. Antimony mining in the South Island last century has left no visible remains.

What really gripped the public's interest were the successive discoveries of gold, beginning at Coromandel in 1852 and followed by rushes in Nelson, Marlborough, Thames, Central Otago, the West Coast and Hauraki. These alluvial finds stimulated increased population, although it was often transient. Eager goldseekers flocked in from California, Australia and China.

After the rushes subsided a more settled form of gold mining took place, with companies having equipment to extract hard rock for the gold. Probably the biggest boost to profitability was the discovery and widespread adoption of the cyanide method of treatment.

Two other forms of gold-winning proved successful. Hydraulic elevation was used mainly in Otago and the West Coast. Water was delivered into pipes with a jet effecting high pressure to wash down alluvial material from the hillsides. Gold-bearing rivers in both of these regions were mined by dredges which also worked the alluvial flats forming their own ponds.

Henry A. Gordon, who was Chief Inspector of Mines in New Zealand, 1882-1896. He was President of The Australasian Institute of Mining & Metallurgy in 1903.

Coal mining from the late 1850s took place on both islands with production from many areas. Other minerals were mined spasmodically on a small scale.

One of the significant effects of mining has been the very considerable boost given to the engineering industry. Foundries and engineering works produced all manner of machinery and equipment such as stamp batteries, berdans, fluming and pipes, hydraulic elevators, steam engines, water wheels, Pelton wheels (made under licence), dredge tailers, stackers and buckets.

Apart from the School of Mines, formerly at Otago University and since 1987 a Department within the Faculty of Engineering at Auckland University, there were 29 goldfield schools from the mid-1880s to improve the technical knowledge of mine staff.

It is the visual evidence of mining seen in the various structures and relics on old sites that particularly interests New Zealanders.

Copper Mining

Virtually nothing exists of early manganese workings but there is some evidence of copper mining. At Miners Point on Kawau Island, a

Mining Department at the Otago School of Mines, which operated between 1891 and 1909.

pumphouse on the Cornish model was built about 1847 of local Mahurangi sandstone with a brick chimney. It is now a picturesque ruin, having been stabilised recently. After the experience of decomposing ore in transit it was decided to build a smelter of sandstone in a cove of Bon Accord Harbour. Also a ruin with no roof and some incomplete walls, it awaits stabilisation.

On the Roding River near Nelson the ruins of two smelters built of stone and brick in 1886 and 1908 can be seen. A tramway brought down serpentine rock containing copper from the Champion Mine. The United Mine on Dun Mountain also produced copper.

Ruins of the copper mine pumphouse, Kawau Island, built about 1847.

Gold Mining

From 1867 onwards the Coromandel Peninsula saw companies being formed with capital to install the machinery needed to extract gold from hard rock. The major item was the stamp battery, usually operated by water. Here the ore rock was crushed by iron-shod wooden stamps or more commonly cast iron and later steel. These fell on to shoes of the same diameter with the stamps alternating in sequence of falling. Sometimes rock was broken down beforehand in rockbreakers or 'grizzlies'. Over the mine shaft a headframe, commonly called a poppet head in New Zealand, was erected to support the winding gear for the cage. Once numbered in many hundreds, only two remain.

In 1889 the Cassel Gold Extraction Co., having acquired the new MacArthur-Forrest cyanide process for removing gold from crushed ores, bought the Crown Mine Battery at Karangahake to assess its worth in the field. The immediate success of the first field test in the world ensured a steady demand for the process and in 1897 the New Zealand Government took over the patent rights to facilitate its adoption through lower royalty payments. Gold mining had been in the doldrums and the new discovery enabled greatly increased gold and silver recovery with a consequent renewal of capital investment and mining confidence. A few mines, notably the Martha at Waihi, produced silver as well as gold.

Concrete B&M cyanide tanks, Union Hill, Waihi, built about 1905.

Early cyanide tanks were wooden but were soon superseded by steel and even concrete. Steel tanks were usually cylindrical and low in height. In 1902 C. F. Brown of the Komata Gold Mines designed and erected the 'tall' tank using agitation. With a height of 11.5 metres and 3 metres diameter and a 55° conical base it had charges of compressed air passing through the potassium cyanide treated slime to speed up the process. Known as B&M (Brown and McMiken) these tanks soon became popular, not only in New Zealand, but in other gold producing countries. In the United States they were called Pachuca tanks. The largest B&M tanks in New Zealand are at Union Hill in Waihi – a cluster of six erected in concrete by the Waihi Gold Mining Co. about 1905.

In 1904 this company, the largest and one of the most progressive in the country, built its No. 5 pumphouse on Martha Hill in Waihi. An imposing concrete building now derelict it housed a massive Hathorn-Davy compound differential engine to operate the 52.8 tonne pump. In 1913 electric centrifugal pumps took over, for in that year the company opened its Horahora power station on the Waikato River. Electric power was brought to a concrete substation at Waikino by a 77 kilometre transmission line.

Waikino, on the Ohinemuri River six kilometres from Waihi, is where the company began building its huge Victoria Battery in 1896. The range of 100 stamps was made by A & G Price's foundry in Thames, who won the contract from worldwide competition. Power came initially from two turbines each of 149 kilowatts using water from the Ohinemuri River. By 1902 a further 100 stamps made this one of the largest batteries anywhere. Eventually this company had 330 stamps in use. Power came from its Horahora hydro-electric station from 1913. At its peak the Waihi Gold Mining Co. employed 1,500 workers.

Waikino is the site of an assemblage of 32 B&M agitator tanks (cyanide) built sometime before 1910. The steel tanks, removed when operations

Rimu dredge in Westland, 1920s. The dredge scooped gold-bearing gravel and sand from the river bed, extracted the gold, then stacked the tailings behind it.

On the Otago goldfields, about 1900. A stone house built by European and Chinese miners at Arrow Falls. The roof is made of tussock matting and sacks, the chimney of five-gallon tins. The first Chinese miners arrived in New Zealand in 1865 and by 1874 almost 5,000 were at work on the goldfields.

ceased in 1952, sat on arched concrete bases which now present intriguing honeycomb forms as an industrial monument.

The Central Otago goldfield was scattered throughout a mainly arid mountainous region with problems of climate and communications. One of several interesting mining operations was the Bullen brothers' Phoenix Mine on Skippers Creek, a tributary of the Shotover River. The quartz reefs were discovered in 1863 and two years later a battery operated by a water wheel was in use. After 20 years it was replaced by a small hydro-electric plant, notable for being the first such plant on a goldfield. Two Pelton wheels 1.8 metres diameter generated electricity transmitted by 8 gauge copper wire over a 250 metre range some three kilometres to the battery. Today, only the concrete foundations remain and a centennial plaque commemorating the achievement.

One can still see abandoned mining relics on exposed and remote sites among the tussock and schist tors of Central Otago's remarkable landscape. There are stamp batteries, water wheels, hydraulic elevators, fluming, water races, mine shafts and drives, stone huts and rock-cave shelters. High up on the Carrick Range the 8 metre diameter water wheel of the Young Australia Mine stands sentinel like. In a gully below is a stone hut and the five stamp battery. Water for this was conveyed in a race some 20 kilometres long over the 1,265 metre Duffers Saddle.

No. 5 pumphouse of the Martha Mine, built 1904.

At the Golden Progress Mine near Oturehua the poppet head is still intact, with rusting boilers and flue nearby. Although mining began here in the 1880s the present structure was built in the early 1930s. At Macetown, accessible only by tramping or four-wheel-drive vehicle, the remains of the Homeward Bound Battery can be seen. Also in this area is another ten-stamper used by the United Goldfields Co. Near remote Serpentine a ten stamp battery has its well-preserved water wheel alongside. Yet another ten-stamper can be seen on the Rise and Shine Creek in the Bendigo (NZ) goldfield. The Come on Time Battery was re-erected here in 1880. There are many other relics not described.

On the West Coast gold mining was often in dense bush with heavy rainfall. This certainly provided an abundance of water power and some very large water wheels were in use. At the Wealth of Nations Mine near Reefton the diameter was 15 metres. It was replaced in 1907 by a Pelton wheel. Sometimes in winter the water wheel was festooned with icicles – dramatic but not productive. Another large wheel, 12 metre diameter overshot, was installed by the Great Western Co. at Kaniere Flat to operate two California pumps. The water supply was brought by a 15.2 metre high flume on timber trestles.

Poppet head at the Big River Mine, before restoration.

Although not involving any mine structures, such areas as Woods Creek and Nelson Creek illustrate the patience and skills of early miners, especially Chinese, in handstacking excavated rock, cutting deep tail races and driving tunnels for diverting water for sluicing. Today the evidence of these can be seen by the public.

Two important sites in the Reefton district are Big River and Waiuta. At the former, mining began in 1887 but the major developments were from the first decade of this century. The 18 metre high poppet head has been restored in recent years. Below the peak on which it stands is the somewhat derelict wooden engine house with the huge brick-encased boiler. Alongside were the carpenter's shop, blacksmith's shop and a bathhouse. On the same side of Big River is the battery, but without its enclosing structure. The ten stamps were operated by a 12 metre diameter water wheel later replaced by a Pelton wheel. Across the river are the shallow-type cyanide tanks erected in 1908. The mine had a gravity aerial ropeway to bring ore to the battery. This is now derelict. The mine complex is of added interest in having its own sawmill and coal mine in the vicinity but higher than the 800 metre elevation of the poppet head. The Big River Mine is probably the most complete complex in the country, despite its dilapidation since closure in 1942.

Water wheel at the Young Australia Mine on the Carrick Range, built about 1875.

Mining at Waiuta began in 1905 with the Blackwater Mine and was followed by the Prohibition Mine at a higher elevation. This reached a depth of 879.5 metres, the deepest shaft in New Zealand. Ore was treated in a large water-powered battery above the Snowy River. Below it are the rusting cyanide tanks, including two B&M agitators. The battery building and machinery have long since been removed. In 1937 a high technology ball mill was built near the Prohibition Mine and used with the flotation process. Some years ago the poppet head, already devoid of its timber sheathing, was totally destroyed by an out of control

Sluicing for gold, Saint Bethens Range, Otago.

gorse fire. Today the remains of the mill can be seen. At Waiuta a steel boiler flue and a shed remain at the Blackwater Mine.

On the West Coast as in Central Otago dredging was a successful method of winning gold. In the 1930s the author remembers White's Electric and the Barrytown dredges working the flats near Barrytown. Relics of earlier dredges can be seen at Kapitea Creek, Gillespies Beach, and a complete structure at Moonlight Creek.

Early dredging began in 1863 on the Clutha River near Millers Flat with the primitive spoon dredge – a leather 'spoon' or bag suspended on a pole – but understandably it was not very successful. It soon gave way to the current-wheel dredge which used the force of the river to drive paddle wheels to work the bucket chain. The real improvement came with the steam dredge. Plentiful cheap supplies of lignite close to the river banks provided fuel. In 1891 the first electric powered dredge was developed and successfully operated on the Shotover River.

Improvements in design and the invention of the tailing stacker gave rise to a boom period when dredging was carried out on various rivers in Otago including Southland as well as on the West Coast. A thriving industry developed in Dunedin, where a number of engineering firms specialised in dredge manufacture. On the West Coast dredges were made by the Dispatch Foundry of Greymouth. International recognition for dredge design and development was an indication of its importance and in the peak year 1903 there were 201 dredges in use, 52 standing idle, 23 under construction, 14 under removal and two that had been wrecked.

Gold was mined in the Nelson Province last century and there are still a few relics including battery remains. Near Cape Terawhiti on Cook

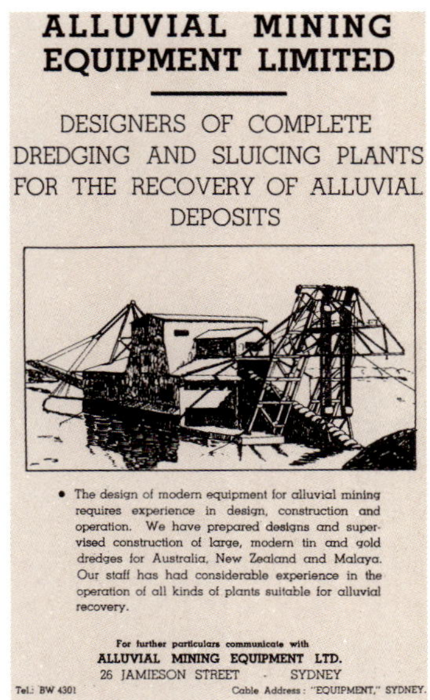

Australian and New Zealand mining people have always had close connections. This advertisement for an Australian manufacturer refers to gold dredges it supplied to New Zealand.

Strait several Wellington capitalists were kept in hope for a time during the 1880s. Today, apart from some adits, all that can be seen are a boiler, berdans, some battery remains and a solitary small concrete store that was also a retort house.

The remotest area in which goldmining took place is Preservation Inlet, at the south-west tip of the South Island. In the 1890s the Golden Site Mining Co. had a ten-stamp battery driven by a Pelton wheel, still visible today in the steep bush-covered gorge. A town called Cromarty was established and another settlement at Te Oneroa served the Morning Star Battery, also located in heavy bush. Today only a camshaft remains.

Boiler house ruin, Big River Mine.

One of several Schools of Mines still standing and the largest on a goldfield is that at Thames, now owned by the New Zealand Historic Places Trust. This agency has a comprehensive management and development plan to put the buildings into a good state of repair and present them in an educational manner for public appreciation. At present only the mineralogical museum is open. The Thames School was renowned for its work on the refractory ore bodies and the difficulties in treatment by the cyanide process. Today the former Coromandel and Waihi Schools of Mines are open as museums. Visitors to Coromandel may also see the former Government Battery operated originally as part of the local School of Mines. The largest school in the South Island is at Reefton, but to date little has been done to present it to the public. Nearby at Blacks Point a former church is now a mining museum with a working battery alongside as an attraction. At Macraes Flat in Otago the Golden Point Mine has a battery now operated periodically by the Department of Conservation as part of the Otago Goldfields Park. Mining began in this district in 1875 and has included some scheelite production. The Otago Goldfields Park came into being about 12 years ago.

Coal Mining

As a comparison with the lure of gold, the public appreciation of coal mining is minimal for it has not the glamour associated with the goldfields. Economic importance makes little impact except as a local source of employment. Nevertheless there are several significant relics of early coal mining and these should be carefully nurtured as evidence of the industry's importance.

Old School of Mines at Thames, now owned by the New Zealand Historic Places Trust.

The oldest coal mine site preserved for public appreciation is the Brunner Mine, 11 kilometres north of Greymouth on the Grey River. Opened in 1864 it is remembered for the underground explosion of 1896 with the loss of 65 lives. Over a period of about 65 years another eight mines were worked in the vicinity. Today there are interesting relics such as boilers on a public walkway tracing the early miners' trail. The most important visual evidence, however, is the presence of the 1868 twin coke ovens and the 1875-90 bank of beehive ovens. Although the latter are partly demolished the public can appreciate the ruins from a covered catwalk protecting the brickwork from foot traffic and damaging frost. The ovens were top loaded with slack coal from a light tramway. Another feature is the massive brick portal of the return airway to the Brunner Mine. Across the river stands the brick chimney of the Tyneside Coal Mine. The New Zealand Historic Places Trust administers this industrial site, which also includes the remains of the associated brickworks. Over recent years it has stabilised the structures and provided a good standard of presentation for the public.

Further north, at Denniston, a bleak plateau 600 metres above the coastal plain became home for miners from 1880. The design and construction of the renowned self-acting Denniston Incline by Henry and Robert Young for the Westport Colliery Co. enabled wagons to transport coal down to the railway at Conns Creek. The minimum gradient was 1 in 1.34. Although after 87 years of operation the track was uplifted the nature of the achievement can still be visualised from the former brakehead area. Over this period 12.5 million tonnes of coal were carried down the incline. The working area behind the brakehead is remarkable for the extent of expertly-laid stone retaining walls where former structures such as workshops, bins and screens once stood on rocky benches. A short distance away is the stone Banbury Arch – a bridge supporting the tramline to the old Banbury Mine. These remains are striking evidence of the work required in building on such an inhospitable site. At the foot of the incline the solitary reminder is the hydraulic crane for handling mine supplies.

The most impressive of the coal mining structures are the large tippling bins and screens. The former bins at Granity were huge. The next largest is at Rewanui, completed in 1913 but, after being badly damaged by a landslip, modified in 1918 by reducing the roof by 4 metres. Serving the Point Elizabeth Coal Mine Co. with four lines of railway track passing

New Zealand Steel plant at Glenbrook produces a range of steel products.

Aluminium smelter at Tiwai Point. The aluminium is smelted in New Zealand from alumina refined in Australia (at Gladstone) from bauxite mined in Australia (at Weipa).

below, it had a storage capacity of 4,080 tonnes. When the Liverpool Mine closed a few years ago it passed into the ownership of the Department of Conservation.

At Kaitangata, also in the South Island, a smaller bin survives. It has steel supports instead of the customary heavy hardwood timbers. In the North Island the impressive steel-framed and brick-panelled Carbonisation Plant at Rotowaro is more recent, dating from 1929-30. Nearby is the fine engine shed built of the same attractive Huntly brick, as also is the mine ventilating shaft alongside.

Aluminium ingot ready for export to Japan at Tiwai Point.

Conservation of the Mining Heritage

Mining is a most important part of the Industrial Revolution which began in Britain in the eighteenth century, and continues to the present day. In New Zealand most of our early mining was small scale, with strong economic pressure to keep investment in buildings and machinery to a modest level. These structures with their plant were often relocated and the effects of very wet climate, exposed sites, lack of maintenance and finally abandonment have taken toll of much of the visual evidence that remains.

Mining has affected all aspects of the lives of New Zealanders over a long period. The New Zealand Historic Places Trust has been active in devoting part of its slender resources into fostering public awareness and an appreciation of our industrial heritage. In the field of mining it owns the Thames School of Mines, Brunner Mine Industrial Site, and as part of the Department of Conservation it is now involved in the Otago Goldfields Park as well as historic reserves such as Denniston. The Trust has statutory responsibility for classifying buildings for their preservation. Many mining structures have been listed and researched to date.

Public interest is stimulated by the former Schools of Mines museums at Coromandel and Waihi. At Westport the Coaltown Museum specialises in Buller district coal mining. The more inaccessible mine sites are visited more frequently by trampers in such areas as Big River and parts of Central Otago. Sadly little, if anything, has been done by the various mining industries or the former Mines Department to help in the conservation of surviving structures that can teach present-day New Zealanders and overseas tourists something of what mining heritage should mean to us as a small, relatively-young but technologically-vibrant nation. In this our sesquicentennial year there is still a challenge to meet.

I am grateful to Geoffrey G. Thornton, Deputy Chairman, New Zealand Historic Places Trust, for permission to include this paper, which he presented to the Annual Conference of The Australasian Institute of Mining and Metallurgy at Rotorua on 20 March 1990.

Fiji

VANUA LEVU
Vatia Point
Tavua
Vatukoula
VITI LEVU
SUVA
0
50
100km

Fijian Gold

Geologists visited the Fijian Islands in the 19th century, but their interest was mainly in the nature and structure of the Pacific Islands. It was not until about 1909 that the prospect of economic mineralisation was regarded seriously and N. D. Cochrane was appointed Mining Adviser to the Fijian Government. His numerous reports noted 'evidence of mineralisation' on the two main islands, Viti Levu and Vanua Levu.

The Yanawai goldfield, in western Vanua Levu, which was worked by Kasi Mines Ltd from about 1931 to 1946, was the first significant gold field in Fiji. Evidence of gold had been known since 1872 in the Tavua Basin area, but economically mineable reefs were not found there until 1932 when Bill Borthwick, an experienced prospector, found Vatukoula, which in Fijian means 'rock of gold'.

Vatukoula had a boom in 1935-36 when more than 20 prospecting and mining groups were active in the region. Of these, only Emperor Gold Mining Company Ltd, Loloma (Fiji) Gold Mines N.L., Fiji Mines Development Ltd (later called Dolphin Mines Ltd), Koroere Gold N.L. and Aloha Central Gold Mines N.L. remained in 1937. Emperor pioneered the Vatukoula field and remained the largest producer. Development of these mines led to provision of the deep water port at Vatia Point.

Borthwick Shaft crew, Vatukoula, 1966. Bill Borthwick, an experienced prospector, found gold at Vatukoula in 1932.

Emperor acquired Koroere in 1938 and Aloha Central ceased operations in 1942. Emperor, Loloma and Dolphin, which had common Directors, were brought under the same general management in 1939 when Nils Nilsen became Chief General Manager and by 1942 custom treatment of ore from the three mines began at the Emperor mill.

The principal producing areas have been the shatter zone at Emperor, the Duchess orebody of the Loloma mine and the several Dolphin Lodes on the Dolphin Leases. A large open-cut was developed at first and operations later extended to greater depths via shafts.

The original workforce consisted mainly of Fijians, some Indians, Europeans, part Europeans and Rotumans. The indigenous peoples showed a remarkable capacity for mineral industry work and became very skilled in geology and mining and milling practices. Many have achieved tertiary qualifications in a range of mining-related professions and hold senior positions.

In an early action in 1945 Emperor introduced a safety campaign to 'instruct and encourage the native miners to work safely'. It was remarkably effective, reducing the lost-time-injury frequency rate from 255 per million manhours in 1945 to 36 in 1951. It has been further reduced since that time to about 11 lost-time accidents per million manhours today.

In 1958 Emperor was the only company still operating and in June 1961 it acquired the remaining assets of Loloma and Dolphin.

In 1966 record rates of development driving were achieved at Vatukoula, when a 2.44 metre by 2.90 metre drive on 11 Level advanced 325 metres in four weeks to 9 March, equal to 11.6 metres per day. Later 400 metres of crosscut and 21 metres of equivalent stripping were completed in 30 working days. These achievements are all the more remarkable when we consider that the driving was done with ordinary air leg equipment and the rock temperature on the 11 Level was 36°, rising to 42° on the 16 Level.

The original pilot plant for treating oxidised ore at Emperor began operating in 1935 using standard cyanidation at a rate of 26,000 tons per annum. It was enlarged in 1937 to 156,000 tons per year and later enlargements brought its capacity to 640,000 tons per year. When the management of Emperor, Loloma and Dolphin was integrated in 1939, Emperor was by far the biggest producer.

Guy L. Ditchburn was Assistant General Manager and Metallurgist when the plant was established. His plans to enlarge the mill to treat all the ore from the Emperor, Loloma and Dolphin mines were disrupted by the Second World War and it was not until October 1944 that combined milling commenced. In addition to flotation, two Edwards type roasters with 60 rabbles were installed, one having come from the Youanmi Mine in Western Australia. The pre-cyanidation conditioning of the ore was based on the successful process developed in Kalgoorlie. A. J. Steel, Mill Superintendent in 1952, described the treatment process in the AusIMM Proceedings in 1953.

Airleg development, Vatukoula, 1966. Record rates of development were achieved at this time.

The founding Directors of the Emperor Mines Ltd were Wallace H. Smith (Chairman), P. F. Cody, Sir Maynard Hedstrom, Edward G. Theodore (Managing Director), T. R. Victor and J. Wren.

Vatukoula's Wallace Smith Shaft is named in honour of Wallace H. Smith, the Melbourne stockbroker who with L. G. May formed a very-successful syndicate known as Australian Gold Development N.L., with major interests in Golden Plateau N.L., Cock's Eldorado Gold Mine Ltd, Emperor Gold Mines and Loloma (Fiji) Gold Mines N.L.

Nils Nilsen, American-born of Norwegian parents, was an Education Science graduate who became General Manager of Emperor Mines Ltd in 1936 and had a great impact on Vatukoula. He became a Director in 1960 and later Chairman of several gold mining companies. On presentation of the Institute Medal to him in 1970 Professor Eric Rudd, AC, then AusIMM President, said of him:

> Over these years he has had many accomplishments to his credit. Some are technical, some in the realm of public relations and all have improved both mining methods and the community life of his employees. Take a few:
>
> - personally responsible for installing the first scraper hoists, both air and electric, underground at Mount Isa in Australia.
> - associated with breaking development records including shaft sinking, driving and rising in Australia and New Zealand.
> - contributed substantially to greater mechanisation and efficiency in underground operations.

Borthwick Shaft winder room, Vatukoula, 1966.

- created in Fiji a first class mining and milling organisation out of Fijians who had never seen a mine.

Today by his effort 2,000 men with 4,500 dependants represent the Emperor Mines in a happy and efficient community. One has only to visit Vatukoula to realise what a colossal task he had and how successfully he accomplished it.

In 1957 Nilsen was succeeded by Fred Watson, who was General Manager when the record-breaking driving and rising performances were achieved.

D. Nils Nilsen (left) accepts The AusIMM Medal in 1970 from the then President of The Institute, Eric Rudd, AC, Professor of Economic Geology at the University of Adelaide.

A good review of the life and conditions at Vatukoula in the early days by R. H. Yarrow of the Emperor Gold Mining Company appears in the *Proceedings of The Australasian Institute of Mining & Metallurgy* No. 168-9 of 1953 under the title 'A General Description of Community Life and Safety Work in Vatukoula, Fiji'.

In an article in *Australian Mining*, in February 1969, Haughton Dunkin, Associate Professor of Mining at The University of Melbourne, described the beginning of a reforestation programme by the Company on the hillside south of the central area, long before 'the environment' became a popular issue. His five articles were reprinted by Emperor Mines Limited as *Mining for Gold in Fiji*.

The Vatukoula ore contains tellurides, and in the early 1970s about two tonnes of ingot tellurium was produced. But, because of unacceptable mercury levels and high costs, the venture was discontinued. In the mid 1970s there was a change in control of the companies and a large portion passed to New Zealand interests.

Like most gold mining activities around the world, the Fijian operations of Emperor and Loloma suffered in the years when gold prices were fixed and production costs rose steadily because of inflation. By 1978 operations at Vatukoula had been reduced from 7,000 tonnes per week to about 3,150 tonnes. The underground workforce fell from 661 to less than 90 and most of the production came from the Loloma open-cut. After a four-month strike, in 1978, there was some rationalisation of underground activities and where two employees had been used for every operation, regardless of its nature, a one-man-per-job policy was introduced, with appropriate financial inducements. This had a great influence on the company's ability to survive.

The workforce fluctuated considerably, peaking at just over 2,400 in the early 1960s, falling to less than 1,000 in the late 1970s. In 1990 it was about 1,400.

A mini-cage, in a water drainage rise, Vatukoula, 1966.

Joint Ventures

In 1983 Emperor formed two joint ventures with Western Mining Corporation. In the Vatukoula Joint Venture WMC took a 10 per cent interest and agreed to manage the Vatukoula Mine. Its interests rose to 20 per cent in 1984. The Tavua Basin Joint Venture was a 50/50 arrangement, covering exploration in the Tavua caldera. This led to the formation of the Tavua Basin Mining Joint Venture, to develop a gold deposit found in the area.

The first WMC-appointed Resident Manager was Jack McDermott, ex Kalgoorlie, followed by Graeme Sauer, Bill Anderson, and Andy Cullum. All contributed new ideas and improvements, which helped restore profitability, and efficiency, and extended the life of Fiji's most important mineral development. Annual production rate rose to 640,000 tonnes per annum of ore, which in 1988-89 yielded a record 142,500 ounces of gold; about three times the production seven years earlier.

Total production to 1990 was about 4,173,000 ounces (130 tonnes) of gold, ranking Vatukoula among the major goldfields in Australasia after Kalgoorlie, Bendigo, Ballarat, Charters Towers and Norseman. At today's prices, the value of the gold produced is about F$2.5 billion. The export

The treatment plant, main haulage shaft, powerhouse and associated services at Vatukoula during the 1960s.

Apprentices in the Vatukoula electrical workshop, 1990.

of gold in 1989, valued at F$26,200,000, accounted for 13.3 per cent of exports and ranked second only to sugar as a commodity export.

After sugar production and tourism, gold mining is the next most important economic activity for Fiji. In terms of value of output per person employed, it is one of Fiji's most rewarding industries. Everybody associated with gold mining can be justly proud of their contribution to Fiji.

Tipping ore into an ore pass, Vatukoula, 1950s.

One should not overlook the assistance of the Fijian Government in support of operations at Vatukoula. As well as some of the infrastructure, the Government provided financial assistance between 1958 and 1960 for a new crushing and ore washing plant at a critical time in the company's life. Emperor had devoted most of its funds to mine development and exploration.

Other than gold at Vatukoula and Yanawai, the only other significant mining operations in Fiji appear to have been mining and beneficiation of copper ore by a Japanese company near Udu Point, on the north coast of Vanua Levu, and mining manganese ore for a time in the west of Viti Levu. Numerous prospective areas have been explored but no mining development followed.

The major contribution to this paper by Gilbert Ralph, MBE, is gratefully acknowledged. I am also indebted to Andy Cullum, Resident Manager at Vatukoula, for his advice and Bill Cornwall, Assistant Manager, for his contribution to the paper, which was presented in Vatukoula on 12 March 1990.

Papua New Guinea

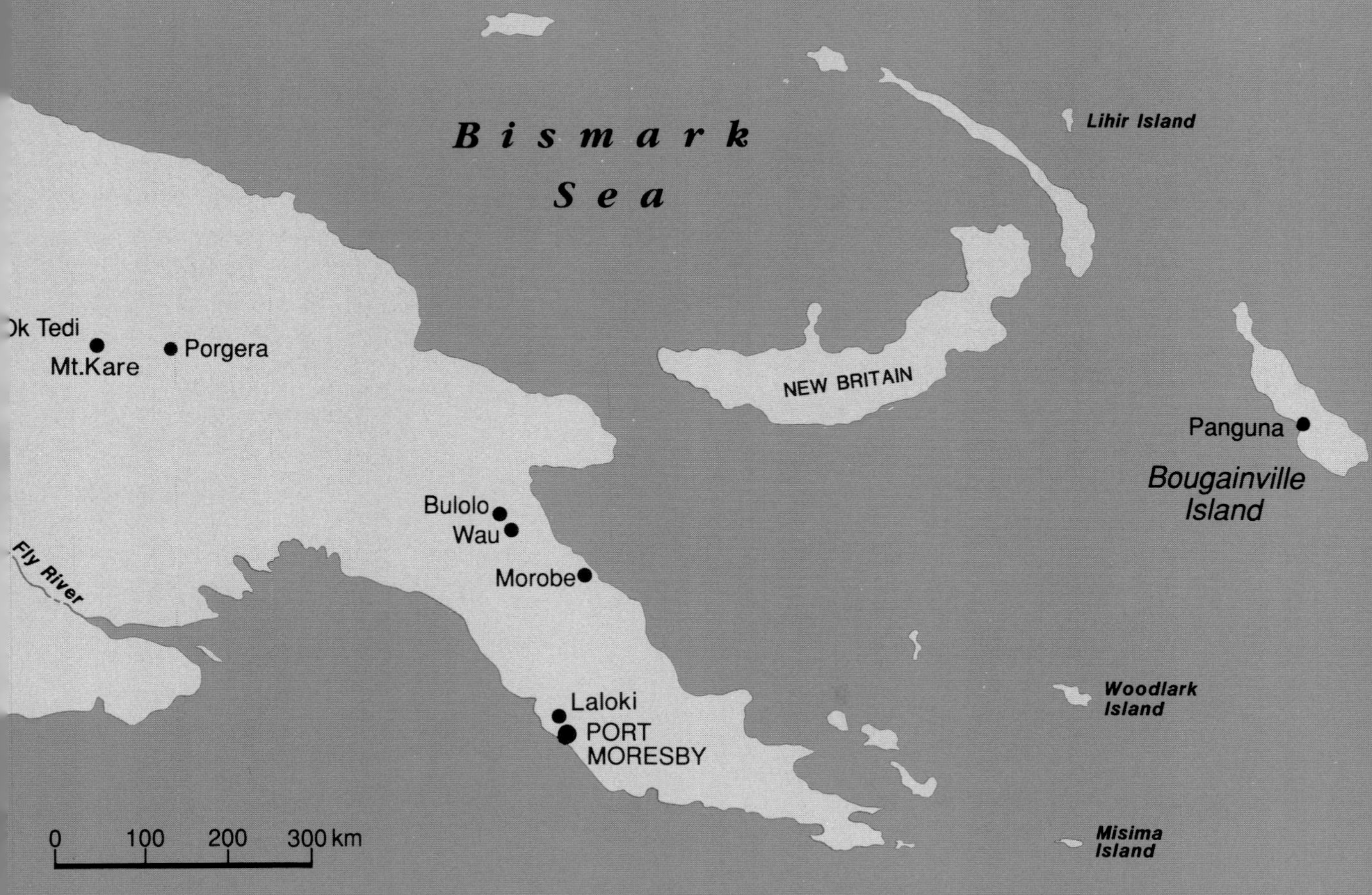

Mineral Wealth in Papua New Guinea

The mineral industry of Papua New Guinea is mainly involved with gold, copper, and petroleum.

Late in the nineteenth century the Queensland Government Geologist predicted that the goldfields of Queensland would extend into Papua. The 100 or so diggers who rushed to the Laloki River near Port Moresby following glowing press reports in 1878 found few colours, but plenty of malaria. Ten years later there followed rushes to the Louisiades. There was indeed gold in these islands and some made fortunes on Sudest and Misima.

The miners soon realised that they could increase their production by hiring teams of Papuans to work the boxes. Papuans soon began to mine for themselves or for the sometimes crooked storekeepers to whom they sold their gold.

In 1895, Charlie Lobb and Dick Ede, traders and part-time prospectors, half filled a billy can with gold from a creek on Woodlark Island. Shortly after, gold was found on the Mambare River, in what is now the Oro Province. These strikes caused gold rushes and marked the start of continuous production to this day.

Happy prospector with 600 ounces of Edie Creek gold in his pan.

In 1896, some alluvial miners formed the British New Guinea Goldfields Proprietary Company Limited to work reefs on Mount Adelaide in the centre of Sudest Island. The Government supported the venture with land grants and prison labour, two commodities it could dispense at little cost. The company bought crushing machinery from Australia, but failed after three years and little production.

In 1899, prospectors found the Ivanhoe lode at Kulumadau on Woodlark Island and promoters, in turn, found the prospectors. The Ivanhoe yielded samples assaying 40 ounces to the ton, a result tailor-made for the stock exchange, and a mine operated until 1918. It was never profitable for the shareholders, but was an important part of the local economy. It employed over 300 men and accounted for about 15 per cent of the exports from Papua.

On Misima Island, lode mining began in a small way before 1910, with increasing attention from 1912. In 1914, O. B. Ward, General Manager of the Broken Hill Proprietary Block 10 Ltd, one of the Collins House group, inspected leases on the Umuna lode. Ward was impressed and arranged the purchase of the mine.

Block 10 could not have moved into gold mining at a worse time. Economic conditions and Commonwealth Government manpower policy during the First World War destroyed the viability of all but the richest gold mines. The industry did not recover until the price of gold increased in the 1930s. The Block 10 company dithered around on Misima until 1922, for very little return.

Early Copper Mining

Copper became of interest around 1910, when prices rose. In 1914, The Great Fitzroy Mines Ltd, which operated the Mount Chalmers copper mine in Queensland, acquired control of the Laloki deposit, but Great Fitzroy collapsed with the downturn of the copper market at the start of the First World War.

The same group formed a new company towards the end of the war, when copper prices were again booming. In 1920 they floated New Guinea Copper Mines Ltd to work the project. This was a major undertaking, especially in the small economy of Papua. It involved opening two mines (Laloki and Debuna), building a railway and a smelter to treat 6,000 tons per month, and installing port facilities. At its height the project employed 100 expatriates and 1,000 Papuans. It promised rapid industrialisation of the Port Moresby area, but it was a complete failure.

What had the mining industry meant to Papua up to this time? In the Annual Report for 1906-07, the Director of Mines, Staniforth Smith, said: 'For the nineteen years of our history, gold mining has constituted practically the only industry – the source of nearly the whole of our revenue. . .' This statement ignored that most of the colony's income was from Australian Government grants, but it was true that in the first 30 years, 1888 to 1918, mining accounted for 62 per cent of all recorded exports.

The mining industry had substantial support from the British New Guinea (and subsequently Papua) Government. Besides providing administration, the Government had allowed prospectors to accompany exploring expeditions, provided access tracks to goldfields (the Kokoda trail was one) and even constructed rest houses on these.

But what was happening in the Territory of New Guinea, north of Papua, which Australia held under a League of Nations mandate after

Dredge at work, Bulolo Goldfields.

a period of German rule that lasted from 1884 to 1914? Almost nothing. The German Administration had little interest in mineral development.

Gold in the Thirties

After the First World War, the Germans had gone and the environment was more favourable for Australian prospectors. Sharkeye Park returned and discovered gold at Wau. As the word got out, he was followed by a trickle of miners. In 1926, Bill Royal climbed into the Upper Edie, found fabulous gold and started a rush. It was the greatest alluvial field in Papua New Guinea until Mount Kare was found. The rush to the field also bred a rush in the cities of Australia. In 1928 and 1929, there was a flurry of flotations of companies to prospect and mine on the Morobe. Ian Morley, ISO, and Charles Marshall, AO, helped form the Morobe Branch of the AusIMM in 1932.

But the real riches were in the flats of the Bulolo. Cecil Levien pegged them for Guinea Gold N.L., a small Adelaide company formed in 1926. They interested Placer Development in the prospect and dredging commenced in March 1932. It was an immediate success. Even before the first dredge was in operation, additional dredges were being planned. In the first three years the company paid over £1,000,000 in dividends, say K20,000,000 in 1990 money. This exceeded the dividends paid by either of the two outstanding Western Australian companies, Lake View and Star and Wiluna Gold Corporation, for the same period. Between 1932 and 1938 Bulolo produced 683,277 ounces of gold and 307,415 ounces of silver from five dredges on the Bulolo and Watut Rivers. The plant did not operate during the Second World War, but dredging resumed afterwards and continued until 1962.

Revolving ball mills at the Panguna copper mine on Bougainville. In these, steel balls grind ore to fine pulp, enabling separation of the minerals from the host rock.

A second major company developed the Morobe goldfield. This was New Guinea Goldfields Ltd, formed in 1929 by Leslie Urquhart of Mount Isa fame. Julius Kruttschnitt was later Chairman. The company continued operations until 1938, but was not as rewarding as Bulolo Gold Dredging.

Papua was severely hit by the Depression, as it depended almost wholly on agriculture at that time. The one bright spot was on Misima Island where Fred Cuthbert had taken over the mine on the Umuna lode. Already an old man, Cuthbert came to Papua from Croydon in Queensland, where he had made a fortune. Some said he could smell gold. Others said he was the meanest man alive, but he could make a gold mine pay. He and his partner bought the mine for £2,000 in 1928. In the first year they produced £6,584 in gold and sold the mine in 1935 for £100,000 in cash and paper. The mine continued to make excellent returns. In 1940 it paid a dividend of £38,730, equal to 20 per cent of its subscribed capital. It was abandoned during the war and not reopened.

In the 1930s, gold provided the impetus and much of the finance for opening up the Highlands. There were many expeditions, those of Jack Hides and the Leahys being well known. Another was that of the Oroville

Panguna copper mine on Bougainville Island became inactive following insurrection in 1989.

Dredging Company Ltd to the head-waters of the Fly and the Sepik in 1935, 1936 and 1937. The Company hired a very experienced team of prospectors, led by the American Ward Williams. An airstrip was constructed at Kiunga, to prospect the alluvials in the Ok Tedi, Fly, Palmer, and Strickland rivers. The results were not encouraging, so a second air strip was built at Telefomin and the Upper Sepik drainage was prospected from there. The search here, too, proved unrewarding and was abandoned in 1937.

After the Second World War, gold mining was slow to recover because of the fixed gold price and inflation. Production by Bulolo Gold Dredging Ltd and New Guinea Goldfields Ltd in the Bulolo and Wau areas continued but there was little other activity.

Copper Since the Sixties

Copper became of interest again in the 1960s. In 1964 CRA's regional exploration geologist Ken Phillips recognised the Panguna porphyry copper deposit on Bougainville Island. The history of this project is well known: the remarkable engineering accomplishments, the problems of land acquisition, the marvellous returns when production commenced, the renegotiation of the Government agreement and, most tragically, the recent insurrection and consequent shutdown of the operation. The Bougainville Copper project has been fundamental to the economic development of Papua New Guinea. Bougainville's annual gold production of about 500,000 ounces was nearly equal to the total Papua gold production between 1888 and 1926.

Petroleum

Oil was discovered in Papua in August 1911 by G. A. Thomas and L. Lett. For political reasons the Australian Commonwealth Government reserved the oil resources for itself and embarked on a vigorous exploration programme. Expectations were high. Evan Stanley, the Government Geologist in Papua, said in 1913 that he hoped a field would be proved or disproved within the next 12 months. It was not to be that easy. In 1919 the Government called in Anglo-Persian Oil Ltd, known today as British Petroleum Ltd, to do the exploration.

The New Guinea territory, where petroleum seeps had been found in 1912, was by then available for exploration as well as Papua. Anglo-Persian did excellent exploration work, but could not find commercial oil. So the Government tried a new approach, though Anglo-Persian continued to operate for some years. From 1923, private exploration was permitted and small Australian companies became active in the field. One of these, Oil Search Ltd, was formed in 1929 and still operates. From the mid-1930s, a consortium of the major oil companies operated through Island Exploration Company and Australasian Petroleum Co. Ltd. After the suspension of operations in 1942, exploration resumed in 1948.

In the years after the Second World War, petroleum exploration was important to the economy of Papua New Guinea, and its personnel comprised a large part of the expatriate community. For instance, in 1955 most of the 578 expatriates and 4,154 Papuans reported to be engaged in surface mineral activities were exploring for petroleum. The only result of this sustained effort was a series of gas discoveries and one short-lived oil flow. Activity died off in the late 1970s, but resurged strongly in the early 1980s as new technology became available.

Meeting in 1990 of Mount Kare landowners, at which they endorsed a joint venture with CRA to develop the alluvials. The Mount Kare gold rush of 1988-89 involved up to 10,000 people.

This led to the further oil and gas discoveries which commenced with Iagifu in 1986. Now, with the Kutubu Project expected to proceed, the reward for the effort of three quarters of a century is in sight. Development of natural gas reserves at Hides is taking place to provide the electric power supply for the Porgera Joint Venture.

Gold Since the Sixties

Taylor and Black noted alluvial gold in the Porgera River during their exploratory patrol of 1938-39. Virtually continuous exploration since the early 1960s eventually bore fruit with the proving of the Porgera deposit, which has an estimated recoverable 12,300,000 ounces of gold. The Porgera Joint Venture began production in July 1990, with gold production expected to exceed 900,000 ounces annually.

On Misima Island, Misima Mines Ltd began gold production in 1989, employing 700 people for an initial annual output of 310,000 ounces. A joint venture was formed by CRA (51 per cent) and the landowners (49 per cent) to develop the alluvials on a Special Mining Lease granted in August 1990. Initial capital was K5,000,000, with production expected to begin in January 1991. In 1990, gold deposits were known at Lihir and Hidden Valley and were awaiting development.

Oil drilling site in Papua New Guinea

Increasing numbers of Papua New Guineans became small scale alluvial miners, replacing the Australian alluvial miner of the pre Second World War period. The Mount Kare gold rush of 1988 and 1989 involved perhaps as many as 10,000 people. One facet of the rush was the extraordinary quantity of gold produced by panning. Another, and as yet unknown, is the long term social effect on those who took part.

Mining and the Economy

The value of Papua New Guinea's exports reached a peak of K1,218,000,000 in 1988 of which K858,000,000, or 70 per cent, was because of minerals, mainly copper and gold. Exploration expenditure in 1990 was about K150,000,000 a year for petroleum and K50,000,000 a year for minerals.

Direct employment in the minerals industry in Papua New Guinea was about of 4,630 persons. Bougainville Copper, before it shut down, employed a further 2,300 people. Employment in the petroleum industry was growing rapidly.

Education and Training

Until recently most of the professionals in the industry came from overseas, mainly Australia, while most of the tradesmen and operators were trained locally. A Department of Mining Engineering, established at the University of Technology in Lae in 1990, had 19 students who expected to graduate in 1992. Before the suspension of operations at Bougainville, over 3,200 people were doing training courses at the mining operations there and at Ok Tedi, Misima and Porgera, and 130 people were engaged as instructors.

The Institute, both directly and as a member of the Mineral Industry Advisory Committee, is assisting in various ways, such as syllabus design.

Mining in PNG and The Institute

In addition to the people already mentioned, many well known Institute members have been active in Papua New Guinea. To mention just a few, Klem Gross, Ian Morley, ISO, Charles Marshall, AO, Lyn Noakes, Norman Fisher, John Hohnen, Jerry Owers, Haddon King, Sir Frank Espie, OBE, and Lord Casey, KG, GCMG, CH, DSO, MC, have left their imprint.

Future Outlook

The minerals industry has been very important to Papua New Guinea. It has made a vital contribution to the economy over the past 100 years and has led to the development of transport and communications throughout the country.

The industry will continue to be essential. Its prospects for further growth are excellent, provided that the political climate and public opinion are favourable and the working environment justifies the vast capital investments necessary.

I am indebted to Bill McGee, a consultant geologist who spent much of his time in PNG, John Dew and Gilbert Ralph, MBE, for their invaluable contributions to this paper, which was presented in Port Moresby on 17 October 1990.

Rabbie Namaliu (centre), Prime Minister of Papua New Guinea, with Sir Arvi Parbo, President of the Australasian Institute of Mining and Metallurgy, and Judy Webber, Chief Executive Officer, in Port Moresby, 1990

Copper and Gold from Ok Tedi

Reports of mineralisation in the Ok Tedi area date back more than a century to 1875, when the D'Albertis and Hargraves expedition penetrated the Fly River and its tributaries to within 50 kilometres of the copper deposit. Hargraves examined sand from the river bank and 'found a speck of gold, also a specimen of copper'. This did not excite much interest at the time. Between 1840 and 1880 more than 1,000 copper deposits were found in Australia and many of them were worked.

For some decades after that, people were rather pessimistic about the economic prospects of the western part of Papua New Guinea. The Oroville Dredging Company Ltd, looking for gold, organised expeditions to the headwaters of the Fly and the Sepik in 1935, 1936 and 1937. An airstrip was constructed at Kiunga, to prospect the alluvials in the Ok Tedi, Fly, Palmer, and Strickland rivers. The results were not encouraging and the search was abandoned.

Early access to the Ok Tedi minesite, in the Star Mountain region of Papua New Guinea, was by helicopter. Early 1980s.

In 1963, almost 100 years after the D'Albertis and Hargraves expedition, Des Fitzer led a government patrol into the Star Mountains and brought back samples of copper mineralisation from Mount Ian, six kilometres north of Mount Fubilan. Three years later, another patrol officer recorded yellow and green encrustation along the entire length of a creek in the area of Mount Ian and presumed them to be sulphur and copper minerals, but these observations passed unnoticed.

It was not until 1968 that geologists with the Kennecott Copper Corporation, using a helicopter in their exploration activities, discovered mineralised float at the confluence of Ok Tedi and Ok Menga and traced this back to its source in the Mount Fubilan region.

Kennecott acted quickly. Only months after completion of its initial exploratory work, the company established a base camp at Tabubil and by February 1969 had started drilling. Kennecott's drilling lasted almost three years, until October 1971. The company also began a number of civic projects, including a medical clinic and a primary boarding school. For the local population, change from then on has been both swift and dramatic.

Negotiations with Authorities

In 1970, the Territory of Papua and New Guinea Administration recognised Ok Tedi as a potential major project and began a series of joint meetings with Kennecott. Discussions continued through 1972. Kennecott wanted some sort of agreement (preferably along the lines of the agreement with Bougainville Copper Limited) before it would commit further funds, but the Administration was unwilling to make a commitment on the available evidence.

Heavy earth moving equipment clearing space for a road to the Ok Tedi minesite, 1982.

Kennecott continued to be active in the field. An engineering report indicated reserves of 137 million tonnes of open-cut copper ore, averaging 0.88 per cent copper and another 100 million tonnes of ore beyond the projected open-cut. The airstrip at Tabubil was completed and preparations for development drilling began.

But Kennecott's development drilling was never to go ahead. At first it was delayed by, of all things, drought. More critically though, the project was later postponed indefinitely when Kennecott decided against further investment. By 1973, the project had been placed on standby. No further technical work was undertaken, although the civic projects continued and most staff members were retained.

A number of significant political developments doubtless influenced Kennecott's decision. In 1971, the Chilean Government expropriated private company operations. In Papua New Guinea rapid change was also evident. As the country moved towards self government, increasingly nationalistic sentiments were being expressed. Bougainville Copper Limited's profitability sparked some unrest and calls for renegotiation of agreements. Kennecott's inability to secure the PNG Administration as a partner in the project contributed to questions about the security of any further investment.

Some of these reservations were laid to rest in 1973 when Papua New Guinea became self-governing. A policy on major mining ventures was adopted and negotiations between the Government and Kennecott resumed in 1973 and 1974. In 1974, legal documents were drafted and the parties announced that agreement was expected in the new year. But early in 1975 more problems arose, and in March it was announced that negotiations had broken down.

The Ok Tedi Development Company

The Government of the newly-independent Papua New Guinea resumed exploration on its own initiative. A government-owned company, the Ok Tedi Development Company, was formed, and many former Kennecott staff joined. Drilling resumed in January 1976.

The Government meanwhile continued discussions with other mining companies. In March 1976, it announced that conditional agreement for further testing and possible development of a gold and copper mine had been reached with Dampier Mining Company, a subsidiary of BHP. The agreement was ratified by the Papua New Guinea Parliament in June 1976 and became the Mining (Ok Tedi) Act of 1976.

Operations in 1990 were still governed by these agreements. The Sixth Supplemental Agreement was then the latest in a number of major modifications. The agreement with the Government was conditional upon BHP forming an international consortium. This was done and an agreement was signed in October 1976, the partners being BHP, Amoco, a group of German companies and the PNG Government.

The then Chief Economist of BHP is on record as saying that 'chance played a part' in BHP's decision to invest in Ok Tedi. The price of gold rose strongly after Kennecott's withdrawal from the project and this made the task of reaching agreement with the Government and finding partners easier.

Throughout the negotiations, which were described as 'complex and protracted', it became apparent to both the Government and BHP that a full scale feasibility study was required. The joint venture set about undertaking such a study. It took over the management and running costs of the government-owned Ok Tedi Development Company, inheriting the administrative facilities in Port Moresby and an experienced team of people who had worked on both Kennecott's and later the Government's drilling programmes.

Late in 1977, Bechtel was appointed as overall coordinating engineering consultant for the project. Drilling crews on Mount Fubilan in those times gave their camps names such as New York, London, and Helsinki. Niagara Falls would have been a more appropriate choice. Site investigations were completed in June 1978. In November 1979, the feasibility study and development proposals were presented to the Papua New Guinea Government and, with some modification, approved. On 21 February 1981 Ok Tedi Mining Limited (OTML) was formed to develop and operate the project.

Ok Tedi minesite, 1984.

In all, 61 kilometres of core drilling were done for the exploration and feasibility study. With no roads in the area, all transport was by helicopter, with the initial investigation and clearing for the helicopter to land being done by hand. The extensive drilling proved ore reserves comprising 51 million tonnes of gold capping ore grading 2.4 grams of gold per tonne, at a cut off of 1 gram per tonne, and 377 million tonnes of porphyry copper ore grading 0.72 per cent copper and 0.5 grams of gold per tonne at a cut off of 0.4 per cent copper. There were also significant probable reserves of skarn ore containing higher copper and gold values.

Development of the Project

The real work began with the formation of OTML. The problems faced during the years of negotiations over the project were mirrored on the technical side. In 1981, Irwin Newman, then General Manager of OTML, predicted that Ok Tedi would be 'the most challenging land project undertaken by BHP since the 1960s.

Geology dictated a three-phase development. Stage One of the project involved the mining and processing of gold ore from Mount Fubilan's enriched cap. Construction began in April 1981, and included a 160 kilometre access road from Kiunga to the mine site, the township of Tabubil, and processing facilities at Folomian. For the first time, BHP had to use an inland river to transport the necessary materials. Some 250,000 tonnes for the Phase One construction reached the area on the Fly River, which a 1982 drought reduced to such a low level that the supply line was halted for nine weeks. Land slides of massive proportions and tremendous geotechnical problems made construction very difficult. Even so, the first ore was fed to the gold processing plant on 15 May 1984, three days earlier than forecast in 1981.

Drilling, Ok Tedi, 1983. The exploration and feasibility study involved 61 kilometres of core drilling.

Stages Two and Three, mining and processing copper ore, involved building facilities to treat copper at Folomian, building a pipeline to carry the concentrate from there to handling facilities at Kiunga, building a hydro-electric power station at Ok Menga, and shipping the concentrate by barge 850 kilometres down the Fly River.

The conditions which made construction so difficult also added to the complexities of environmental management, and several 'incidents' received wide publicity. OTML conducted a comprehensive environmental study to identify the impact of discharge from the mine into the Fly River system. After reviewing the results of the study, the Government set long term environmental standards for the operation.

The three stages of construction cost US$1.4 billion. Mining and processing of copper ore began in 1987 and by 1990 had increased to a rate of more than 70,000 tonnes per day. Processing of gold-only ore ceased in 1988. Concentrate production capacity stands at about 500,000 tonnes of concentrate per annum, containing about 150,000 tonnes of copper and about 350,000 ounces of gold.

Copper and gold are Papua New Guinea's largest export earners, and output from the mine accounted for about 40 per cent of the country's export earnings in 1989. In 1990, the Government expected to earn about K30 million in royalties and other revenue.

Bridge across the Ok Menga River, 1982.

Associated Developments

Mining developments in young countries such as Papua New Guinea are often controversial. Ok Tedi is no different, but the project's impact has been quite positive in many ways. One very pleasing result has been the dramatic improvement in local community health. The infant mortality rate dropped from 30 per cent in 1982 to less than 8 per cent in 1990. Greater control of diseases such as malaria, tuberculosis and filariasis has been another welcome development. Education to high school level became available in Tabubil and a university centre was opened. In 1990, about 500 students were enrolled at community and high school levels.

The Ok Tedi/Fly River Development Trust was established to assist the riverine communities develop their infrastructure, with K2,500,000 made available for 1989. This sum was indexed. Added benefits arose from direct and indirect employment provided by the operation. Some 2,000 nationals and expatriate workers were employed directly, another 1,500-2,000 people indirectly as contractors and the like. Local landowners also received payments in recognition of mining activities and benefitted from the presence of roads, townships and other facilities in the area. The operation attracted people from outside the district. This in turn resulted in increased stress being placed on the services provided by OTML. Increases in local population levels were likely to continue for some years.

These developments were a great credit to all those involved and resulted in very sound relations between the miners and the local community. It is important that this continues and that all parties continue to see the benefits from mining in the area.

Future Outlook

High copper prices in 1990 helped OTML's performance and its debt was being reduced to more manageable levels. At 1990 production rates, the mine was expected to continue operation for another 18 years.

For the local people, the dramatic improvements in community health, improved access to education, and experience gained in running businesses would ensure an enduring positive legacy of mining activities at Ok Tedi.

I am indebted to Vicki Lukins and Steve Harvey of BHP for the preparation of this paper, presented at Tabubil on 18 October 1990.

Ok Tedi processing plant, 1984.

Index

(Numbers in italics refer to illustrations)

Illustration Sources

The publishers gratefully acknowledge the following sources of illustrations.

If only a page number is given, all illustrations from that page are from the same source. Otherwise t=top, c=centre, b=bottom.

Alcan Australia Ltd: 27b.
Alcoa of Australia Ltd: 8b, 20b, 67b, 68t, 186.
Australian Consolidated Minerals Ltd: 145c, 162, 163, 166t.
Australasian Institute of Mining & Metallurgy: 61, 93, 128b, 196b, 197c, 198b, 199, 200, 214b, 247t, 261t, 263b, 276, 288b, 298.
Australian Mining Industry Council: 60.
Battye Library (Western Australia): 153b, 192c.
Bendigo Trust: 242t.
BHP Limited Archives: 25b, 116, 117, 118, 119, 126, 141c, 219, 221, 222, 223, 224, 225, 228, 234b, 258, 299, 300, 301, 302, 303, 304.
BP Australia Ltd: 71t, 297.
Brian Carroll Collection: iv, 2b, 3t, 5b, 6, 8t, 10, 11, 12, 13, 14, 15, 20t, 21, 22, 23, 24, 25t, 26, 28b, 30b, 31, 32, 33, 34, 35, 43, 44, 45, 46, 48t, 49, 50, 52, 53, 54, 57t, 58t, 64, 66, 67t, 71b, 73, 74t, 79, 80, 81b, 83, 84, 86, 94t, 96, 97b, 98t, 100t, 101b, 102, 104t, 106, 108, 110, 111t, 112, 114, 115, 124t, 127, 128t, 129, 130, 138, 142, 143, 148t, 149b, 160, 161, 164, 170, 171, 172, 173, 174, 176, 177t, 201, 202, 203t, 204, 207t, 208t, 210t, 211t, 212, 213, 214t, 215, 217, 220, 227, 230, 231, 232, 233t, 234t, 235b, 239, 240, 241, 242b, 243b, 244t, 248t, 249, 250, 251, 252, 253, 257, 260, 261b, 262b, 267, 268, 269, 270, 271t, 272b.
Brookes, Mrs J. D.: 123.
Capricornia College: 74b, 75, 76, 77.
Chamber of Mines, Metals and Extractive Industries (NSW): 69c.
Cobar Museum: 37, 38b, 39t.
Collie Museum Library: 181.
Council of The School of Mines and Industries Ballarat Limited: 1980, 191t, 243t.
CRA Limited: 28t, 29, 30t, 36, 40, 41b, 42, 69t, 69b, 78, 82, 85, 88, 178, 283b, 284t, 294, 295, 296.
Cyprus Gold Australia Corporation: 105b.
Dominion Mining Limited: 121, 122, 131.
Electricity Commission of New South Wales: 27t, 145t.
Emperor Mines Limited: 264b, 286, 287, 288t, 289, 290b.
Energy Resources of Australia Ltd: 69b, 124b, 125b, 146b.
Gwalia Minerals N.L.: 184b..
Hore-Lacey, Ian: 105t.
James Cook University of North Queensland: 192b.
John Oxley Library (Queensland): 65b, 94b, 95, 98b, 99, 100b, 263t.
Kalgoorlie Miner: 153t.
Lewis, R. B.: 193b.
Mount Isa Mines Limited: 5t, 18, 70, 81t, 87b, 89, 91, 103, 104b, 107, 109, 111b, 144b, 145b, 146t, 265b.
Museum of the Goldfields (Kalgoorlie): Endpapers.
National Library of Australia: 1, 2t, 97t.
NSW Department of Mineral Resources – Historical Photographic Collection: Front Cover (dust jacket), 38t, 39b, 41t, 56t.
New Zealand Department of Commerce: 278b, 279t, 281t.
North Broken Hill Peko Ltd: 17, 271b, 272t.
Northern Territory State Library: 56b.
Northern Territory Department of Mines & Energy: 59b, 125t, 132, 133, 134, 135, 136.
Pasminco Limited: 216b.
Queensland Energy Commission: 87t.
Queensland Mines & Energy Department: 90.
Ralph, Gilbert M.: 120, 192t, 193t, 210b, 218, 226, 282b, 283t.
Renison Goldfields Consolidated Ltd: 274.
Robe River Mining Co Pty Ltd: 175.
Royal Melbourne Institute of Technology: 191b.
Sagasco Resources: 197t, 197b, 198t.
South Australia Department of Mines and Energy: 3b, 4t, 55t, 141b, 203b, 205, 206t, 207b.
Sovereign Hill (Ballarat): vi.
State Electricity Commission of Victoria: 254, 255, 256.
Tasmanian Department of Mines: 262t, 273.
Thornton, Geoffrey (New Zealand Historic Places Trust): 4b, 277b, 278t, 279b, 280, 282t.
University of Auckland: 195b, 277t.
University of Melbourne Archives, The: 196t. Broken Hill South Collection: 47, 51, 144t, 237b. J.W.H. Johns Collection: 292, 293. Mount Morgan Collection: 72. BHAS Collection: 216t, 244b.
University of New South Wales: 140, 141t, 194b, 195t.
Western Australian Department of Mines: 55b, 58b, 169, 180, 182.
Westrail: 183, 184t.
Woodside Petroleum Pty Ltd: 177b, 179.
Worsley Alumina Pty Ltd: 187.
Western Mining Corporation Limited Collection: 7, 16, 19, 48b, 57b, 62, 65t, 92, 101t, 139, 147, 148b, 149t, 149c, 150, 151, 152, 153t, 153c, 154, 155, 156, 157, 158, 159, 165, 166b, 167, 168, 185, 194t, 206b, 208b, 209, 211b, 235t, 236, 237t, 238, 244b, 245, 246, 247b, 248b, 264t, 265t, 266, 281b, 290t.